(REF 29500) £6.60

THE CORNISH MI[NE]

PAST PERFORMANCE A[ND] [F]U[T]URE PROSPECT

J.H. Trounson

THE CORNISH MINERAL INDUSTRY:
PAST PERFORMANCE AND FUTURE PROSPECT

A Personal View 1937–1951

John H. Trounson

Edited by Roger Burt and Peter Waite

The University of Exeter
in association with
The National Association of
Mining History Organisations

First Published 1989 by the University of Exeter
in association with The National Association
of Mining History Organisations

©1989 *The Mining Magazine* and *Sands, Clays and Minerals*

Other publications by J.H.Trounson

Historic Cornish Mining Scenes at Surface (Bradford Barton, 1968)
Mining in Cornwall 1850-1960 (Moorland, 1980-1) 2 Volumes

ISBN 0 85989 334 0

Printed in Great Britain by BPCC Wheatons Ltd, Exeter

CONTENTS

List of Graphs and Maps
List of Tables
List of References to the Original Articles

Foreword	vii
Introduction	xi
Section 1: Cornwall's Undeveloped Mineral Resources in the 1930s	1
Section 2: Cornish Mining in the Second World War	
I The Industry at the Beginning of the War	17
II Past Production and Suggestions for the Future	31
Section 3: Regional Survey of the Mining Districts	
I The St Just and St Ives Mining Districts	51
II The Marazion District	73
III The Wheal Vor or Breage District	91
IV The St Erth, Gwinear and Crowan Districts	103
V The Wendron District	127
Section 4: Some Useful Prospects for the Future	
I Levant to South Condurrow	147
II Tolgus Tunnel to Lambriggan	163
III St Austell-Par Mines to Trebartha-Lemarne	181
Mine Index	191

LIST OF GRAPHS AND MAPS

Fig.1. Diagram Showing the Output of Metallic
Tin and Copper from Cornwall, Estimated
from the Public Sales of Ore   35

Fig.2. Sketch Map Showing the Situations of
the Principal Mining Districts
in Cornwall and South Devon   50

Fig.3. Sketch Map Showing the Situations of
the Principal Mining Districts
in Cornwall and South Devon   146

LIST OF TABLES

Table 1. Output of Individual Mines   36

Table 2. Major Producers of Tin and Copper
Ores   37

List of References to the Original Articles

"The Undeveloped Mineral Resources of Cornwall" *Sands, Clays and Minerals*, Vol.III No.2 (September 1937), pp.99–107

"The Cornish Mineral Industry. Part 1" *Mining Magazine*, Vol.LXVI No.2 (February 1942), pp.47–52

"The Cornish Mineral Industry. Part 2" *Mining Magazine*, Vol.LXVI No.5 (May 1942), pp.195–205

"The Cornish Mineral Industry. Part 3. The St Just and St Ives Mining Districts" *Mining Magazine*, Vol.LXVII No.3 (September 1942), pp.119–130

"The Cornish Mineral Industry. Part 4 No.1. The Marazion, Wheal Vor, St Erth, Gwinear and Crowan Districts" *Mining Magazine*, Vol.LXVIII No.1 (January 1943), pp.9–18

"The Cornish Mineral Industry. Part 4 No.2. The Marazion, Wheal Vor, St Erth, Gwinear and Crowan Districts" *Mining Magazine*, Vol.LXVIII No.2 (February 1943), pp.78–85

"The Cornish Mineral Industry. Part 4 No.3. The Marazion, Wheal Vor, St Erth, Gwinear and Crowan Districts" *Mining Magazine*, Vol.LXIX No.6 (December 1943), pp.329–342

"The Cornish Mineral Industry. Part 5. The Wendron District" *Mining Magazine*, Vol.LXXII No.2 (February 1945), pp.73–83

"Some Useful Prospects in Cornwall. Part 1" *Mining Magazine*, Vol.LXXXIV No.2 (February 1951), pp.73–84

"Some Useful Prospects in Cornwall Part 2" *Mining Magazine*, Vol.LXXXIV No.3 (March 1951), pp.139–149

"Some Useful Prospect in Cornwall Part 3" *Mining Magazine*, Vol.LXXXIV No.4 (April 1951), pp.209–216

FOREWORD

John Hubert Trounson died in July 1987, when he was almost 82 years of age. During the last twenty-five years of his life he came to occupy an almost unique place in Cornish mining. Although never a mine agent, nor a mine manager, his contribution to the industry was probably as great as any of his contemporaries. Even before he left South Crofty, where he was Chief Surveyor, to become Project and Development Engineer at the newly created Camborne Tin Ltd., in 1961, he had begun to assume the unofficial role of spokesman for the Cornish tin industry. Within a decade many saw Jack Trounson, as he was almost universally known, as the Grand Old Man of Cornish mining.

Jack Trounson was born in 1905 in Redruth. His family were long-established millers and tradesmen in the town, and his early life was comfortable and secure. As a youngster his twin interests were mining and steam engines, the former dating from a childhood visit to Dolcoath and the latter fuelled by his proximity to the family's steam lorries. Alongside his deeply held religious and humanitarian convictions, these interests were to remain with him throughout his long life.

During his early years, Jack's family enjoyed considerable financial prosperity and a period at Truro College was followed by an expensive public school education at Mill Hill, Hendon, a methodist foundation. However, his family's circumstances appear to have changed significantly in the early 1920s and it became clear that he would need to return home and seek to earn his own living. To acquire practical qualifications he took up a place at the Camborne School of Mines, graduating in 1926. His first job was as a sampler at East Pool, and within a year he was promoted to assistant surveyor. In 1930, during the tin slump that closed all Cornish mines except East Pool, he undertook various tasks at the mine, sometimes working on surface and sometimes underground. Throughout the 1930s he gained experience, becoming a proficient mine surveyor, and in 1938 he took a job as assistant surveyor at South Crofty, by then the healthiest mine in the county. He remained at Crofty until 1961, becoming Chief Surveyor and a highly respected member of staff. It was whilst at Crofty that Jack expanded his responsibilities to include various aspects of engineering. He designed the layout of the pumping engine and

Foreword

pitwork at Castle-an-Dinas wolfram mine, and supervised its installation. He designed hoisting arrangements and chimney at Palmers Shaft, South Crofty. He undertook the planning and supervision of extensions to the adit system at Dolcoath, Roskear, Cooks Kitchen and Barncoose. It was also during his time at South Crofty that Jack became involved in what he came to see as his major contribution to the Cornish mining industry, investigating and reporting on potential mineral sources in the South-West.

Jack Trounson's life-long concern for the continuation of a healthy mining industry in Cornwall, together with his other abiding interests, were reflected in the many and varied organisations that he enthusiastically supported. When still in his twenties he became involved in moves to save Levant whim engine, which led to the formation of the Cornish Engine Preservation Society, the forerunner to the Trevithick Society. He became its first Honorary Curator, eventually becoming the Society's Life President. At the end of the War he was Treasurer of the Cornish Institute of Engineers, a post he held for many years. With the formation of the Cornish Mining Development Association, in 1948, Jack became Vice-Chairman, becoming Chairman a year later and Honorary Life Member in 1986. When the Cornish Chamber of Mines was re-formed he was elected to its Council. He was President of Cornwall Railway Society; President, in 1970, of the national Road Locomotive Society; was a Member of the Institute of Mining and Metallurgy and a Chartered Engineer. He was honoured with the West Medal by the Cornish Institute of Engineers, appointed consultant to South Crofty, made a Governor of Camborne School of Mines, and eventually, in 1971, was honoured by the Queen with an M.B.E.

When Siamese Tin, Tehidy Minerals and Union Corporation decided to initiate a serious search for minerals in Cornwall, in 1961, it was to Jack Trounson that they turned to spear-head the search. For the next ten years, in his capacity of Projects and Development Engineer for Camborne Tin Ltd., he utilised his vast knowledge of the mining districts to promote mineral exploration on a scale not seen since before the Great War. It was the consequent compilation of reports on mineral prospects throughout Devon and Cornwall that he considered his greatest contribution to the continuance of mining in Cornwall. From July 1940, when the Mining Magazine published his article on the wolfram prospect at Castle-an-Dinas, until his final report in the mid-1980s,

Foreword

Jack produced reports on mines all over the South-West. Some have estimated the total mines reported as high as eighty, and no significant mining district or location appears to have been neglected by him.

The most important assessment of Jack's contribution is that of the experts who used his many reports either for research or for indicating potential mineral sources. Indeed, no serious Cornish mining researcher or exploration geologist during the last three decades could have avoided consulting Jack on some aspect of mining. Mine managers, like G.C. Pengilly and J. Symons, and exploration geologists, like Professor K.F.G. Hosking, of Camborne School of Mines, have identified the areas of Jack's greatest worth. First, he kept the flag flying for Cornish mining. Never did he waver in his conviction that the industry had a long-term future. He would write any number of letters, visit any amount of people or research any aspect of mining to keep his belief before mining companies, the press and government agencies. Second, was his tireless gathering of information on mines and mining districts in Devon and Cornwall. G.C. Pengilly has commented that Jack's encyclopedic knowledge of mine sites was without equal. Professor Hosking believes that the third important feature was his willingness to help anyone interested in mining. Academics and company executives, local researchers and students, geologists and mine managers, all found Jack Trounson completely approachable and always willing to spend time explaining all he knew of the subject queried.

During the twenty-five years that I knew Jack I always found him willing to supply the answer to a question or help identify where the answer might be found. When he was past eighty years of age, Jack drove into South Crofty to assist with my research into the adit systems of Dolcoath and Roskear. His mind was still clear and his explanations lucid. Typically, his infectious enthusiasm for mining was that of a young man, introduced to the subject for the first time. It was that youthful enthusiasm, for steam engines, for his religious and humanitarian beliefs, and for Cornish mining, for which he will be best and longest remembered.

J.A.Buckley
Pool 1989

INTRODUCTION

Neither of the editors of this collection ever met Jack Trounson. We knew of him but we did not know him. To us, like so many others, he had a legendary first hand knowledge of Cornish mining. A lifetime spent in and around the mines of his native county, together with a keen mind and an insatiable appetite for every detail of past working, had given him an unparalleled stock of information and material on the industry which had dominated the south western region until recent times. For as many years as we have been involved in mining history and the discussion of its more knotty problems, we have been told, "if you want to know the answer to that question, you'd best ask Jack. He's sure to be able to help". And we did. On many occasions he provided invaluable help and guidance by telephone and letter. Sadly, however, he is now dead. Not only have we lost the opportunity finally to meet the great man but, with selfish alarm, we realised that we had also lost access to his accumulated stock of knowledge and experience. No one can be so confident of their own abilities that they can easily accept the loss of such an eminent authority. What could be done to salvage at least part of Jack's "data banks"?

It was this question that first prompted us to review his published work. At first sight, the prospect was not encouraging. A review of his output during the last twenty years or so revealed only a list of odd light articles and a picture book or two, which are pleasantly diverting but no real testament to the contribution of this man. We began to suspect that like many others with something of real value to say, Jack had neither the time nor opportunity to set it down. The renown of his treasure chest of information had produced constant disrupting calls on his time. Together with an unstinted support for many other local interests, he rarely had time left to put pen to paper. Certainly compared with A.K. Hamilton Jenkin and D.B. Bradford Barton, the other co-founders of modern Cornish mining history, he had published very little since the 1950s. It began to appear that Jack's recent contribution was more by word of mouth, advice, discussion and suggestion than the written word.

Just when we were about to give up, we came across the series of articles collected here. Published between the mid-1930s and the early

Introduction

1950s, when Jack was still a relatively young man, they are among the most perceptive and enlightening documents on the past, present and future of the Cornish mining industry written this century. It would not be too much to claim that, taken together, they represent the most wide ranging and comprehensive review of Cornish mining published between J.H. Collins, *Observations on the West of England Mining Region* and H.G. Dines, *The Metalliferous Mining Region of South West England*. Certainly they withstand comparison with Hamilton Jenkins far better known series on *The Mines and Miners of Cornwall* and recommend themselves for careful use alongside these much consulted reference works. Appearing in the pages of *Sand, Clays and Minerals* and *The Mining Magazine*, the original articles are not easy to find or use. They have remained largely undiscovered by mining historians and, indeed, by prospecting mining companies, to whom they were originally primarily directed. It is hoped that this new edition will not only make them more widely available but stand as a tangible testament to a man who helped so many.

Although arranged here in four sections, the material in this collection really falls into three parts, reflecting the dates of publication of the original articles. Section 1 was written in the mid-1930s, Sections 2 and 3 during the Second World War and Section 4 in the early 1950s. Although widely spaced in time, the motivation for writing the articles was the same in every case — viz. a substantial desire to revive investment and mining activity in Cornwall. Jack was not one who naively believed that every abandoned hole in the ground still had a pot of gold buried in it. He knew very well that this was certainly not true. He was convinced, however, that there were still many mine sites that had been abandoned prematurely and/or had been insufficiently investigated. Careful investment and good management in these mines would not only return a sound profit to the speculators but help to bring much needed employment back to his native county. Every time there was a change in the general economic climate that made the regeneration of interest in tin mining more likely, Jack took up his pen. Indeed the chronology of his publications looks almost like a derived index of mid-20th century tin prices!

After a fairly active period in the 1920s, south western metal mining came to an almost complete standstill in the depression of the early 1930s. A noticeable improvement in metal prices in the mid-30s

Introduction

prompted the publication of the first article in *Sand, Clays and Minerals*. The wartime emergency presented an even better opportunity. Overseas supplies of strategic metals were vulnerable from the outset and throughout Britain feverish activity took place in the search for alternative domestic supplies. When Malaya, the allies' main source of domestic tin, was overrun by the Japanese in 1942, Jack realised that there would never be a better opportunity for publicising the best surviving prospects in Cornwall. Unfortunately the war, and the briefly revived domestic interest, came to an end before much progress could be made and indeed, before Jack was able to complete his detailed series of regional survey articles. The articles on the mines of central and east Cornwall, promised at the end of the survey of the Wendron district, never did appear. The immediate post-war years saw most interest again being directed overseas but the Korean War, re-armament and the creation of stock piles of strategic metals by the USA and elsewhere, forced metal prices up in the late 1940s. The price of tin on the London Metal Exchange more than doubled between 1947 and 1951 to over £1,000 per ton, or more than ten times its level when the Cornish industry began its sharp decline in the closing years of the nineteenth century. It is against this background that Jack launched the Cornish Mining Development Association and wrote the last group of articles. It is not widely known, but he returned to the same theme again in the buoyant market of the 1970s, when he was commissioned by the Department of Trade and Industry to write a further series of reports on the principal Cornish mining districts. Unfortunately, this work, which built on and expanded the material presented here, was never published. This misfortune is compounded by the fact that Jack himself reportedly regarded them as his very best work. It is to be hoped that they might finally see the light of day at some future date.

The format of every section, or group of articles, was rather similar. They started with general observations on the mineralisation of Cornwall and went on to conduct relatively detailed surveys of particular mining districts and mine sites. Although the mine reports will probably receive most attention from researchers, the general comments are more important for our understanding of Jack's overall approach to mining and particularly the "predictive" theories which he thought most accurate and useful. Section 1 effectively sets out the parameters of the study that he undertook in greater detail later. Firstly, the centre of

Introduction

interest was to be firmly focused on metallic minerals — "earthy substances like China Clay and various types of stone which, in the popular sense of the word are not minerals" were entirely excluded. Secondly, as in any proper study of the economics of metal mining, he starts with an overview of the geological features that created the basic parameters and conditions for the industry. He sketches a simple, perhaps oversimple, view of the geological history of the South West and provides a succinct statement of the intrusive/zonal arrangement model of mineralisation — viz. a process of granite intrusion into overlying sedimentary "killas" strata, which had a metamorphosing effect on the killas immediately adjacent to the granite. "The greatest area of stress and fissuring occurred at or close to the margin of the granite intrusions, in other words, in the metamorphic aureole...and in the outer 'skin' of the granite itself. And it is just in or close to this granite-killas junction that the overwhelming bulk of the economically important minerals occur...in general terms it is noticeable that the metallic minerals deposited in the veins are in a definite sequence or zonal arrangement". Tin oxide occurs in or close to the granite, giving way to tungsten ores and then arsenic and finally copper in the killas. Tin can often be found throughout this arrangement but tends to decrease with increasing distance from the granite. With gradual progression through the killas, copper tends to die out and is succeeded in upward order by ores of zinc, lead, iron and manganese. Over time, much of the upper levels of the killas had been weathered away, leaving only the lower sections of these original arrangements. In some areas it might now be the copper exposed at surface while in others denudation continued right down to tin in the granite.

Modern geologists have little time for such simplified theories but, as will be seen at several points in this collection, Jack had even less time for geologists. All that he was interested in was whether there was good empirical evidence for the theory and whether it had practical predictive use. As far as he was concerned, the answer to such questions was clearly yes, though with certain important restrictions and constraints. He accepted that it had been shown that the model was not everywhere applicable to Cornwall. The arrangement of copper and tin ores was certainly reversed in some districts, with tin being obtained from the killas and copper from the granite. Also it had been shown on many occasions that simply to get close to the margin of the granite, *no matter at what depth*, did not ensure discovery of mineral bearing veins. "There are now good grounds for thinking that the bulk of the economically valuable ores in Cornwall lie relatively close to the margin of the granite bosses at present exposed". Even in these areas, only a very small percentage of the total margin was mineralised. This latter observation leads him to the very significant conclusion that, "in the main, the search for unworked deposits of ore is brought back to the old and existing mining districts, in or close to the margin of the great intrusions".

The question to which this whole series of articles was addressed is "whether there is any further prospect in these areas, particularly in regards to tin, and whether previous generations have not completely exhausted them". Unlike some, Jack's advice here was careful and measured. He had a very high regard for the intelligence and abilities of the "old men" who were "just as astute, and far more capable, than present-day mining men in the art of working successfully relatively small and very complicated mineral deposits such as constitute the majority of Cornish mines". Neither did he see the "old men" at any serious technological disadvantage compared with the modern miner. "It is an oft repeated fallacy that the "old men" could not concentrate the ores as successfully or cheaply as can be done today...with the exception of the invention of the magnetic separator... we are probably no better and no worse than the old men as far as concentration is concerned". With such attitudes in mind, Jack counselled strongly against major re-investment in mines that had previously been worked specifically for tin, *unless there was convincing evidence that they were prematurely closed and that there was good prospect for future profit.* If such clear evidence was not available, it should be assumed that the "old men" had found all of the profitable ore and had worked it as efficiently as any modern company. Instead, Jack advocated the re-examination of "shallow copper mines in killas rock which were closed down on the exhaustion of the copper at a time when it was not generally realised that tin was frequently to be found in even greater quantities below the copper". Even then, care should be taken to select mines which were in killas near to the granite in depth and in districts renowned for both tin and copper production.

After these opening remarks, the greater part of Section 1 is taken up with a review of, firstly, the copper districts where new discoveries of tin might be expected at depth and, secondly, specific tin mines which were once worked at a profit but where closure could be considered

Introduction

premature and in no way the result of the impoverishment or exhaustion of the tin lodes. In this review Jack frequently demonstrates the depths and value of his own long association with mines throughout the county. In preparing the re-investigation of Wheal Vor, for example, he is able to dismiss easily the failure of the last attempted re-opening as of no real importance. Who else would know that "the unsuccessful attempt to unwater Wheal Vor about 30 years ago only reflects on the machinery and methods employed at that time"? Who else could confidently conclude that "The water was only partially pumped out and the prospects remain as unproved as if nothing whatsoever had been done". This collection as a whole abounds with such examples of hitherto unknown and invaluable material, not just for the historian but also for future prospecting mining companies. Perhaps the greatest testament to its overall reliability and value is the now proven success of some of his predictions. Many of the prospects that he suggested that have been re-investigated have proved profitable — what would be the success rate for the remainder?

Section 1 concludes with a survey of metals other than tin that might still be found in commercial quantities in Cornwall, for example, arsenic, tungsten, copper, lead, zinc, iron, manganese and uranium. Clearly some of these metals, such as tungsten and arsenic, which were often found in association with tin, might provide profitable by-products for tin mines in the future but it was unlikely that the deposits would ever be large enough to sustain continued mining activity in their own right. For copper, lead, zinc, manganese and iron he notices that these metals had usually been raised from shallow deposits in the past and concludes pessimistically that they were therefore probably all worked out. Jack's clear view was that "tin, with its associated arsenic and tungsten ores, is the only metal which is likely still to exist over wide areas and in really large quantities, and is so situated as to be mined successfully in the future". The only major exception to this rule was the possibility of finding a range of minerals at depth in the Great Perran Iron lode. Jack was firmly of the view that this deposit would repay further investigation. To him it was not simply a shallow worked out iron deposit but probably "a great 'gozzan' indicator of an exceedingly important copper and tin lode at greater depths", possibly also containing valuable quantities of several other ores. A brief diamond drilling exploration in the 1960s failed to prove this prediction but neither did it clearly dismiss it.

Section 2, drawn from a series of articles started in 1942, starts

Introduction

with a brief review of the consequences of the war for the supply of strategic metals to British industry. Here, unlike in the earlier material, Jack is no longer concerned with attracting *commercial attention* to the surviving prospects for *profitable* mining in Cornwall but simply with indicating where and how essential increases in output could be obtained. The imperative need now was "to obtain from Cornwall the maximum amount of tin, tungsten, arsenic and other vital metals" and the essential question in reviving the workings was "one of supply and time rather than economics. In other words, what could Cornwall produce within a limited period." In a forthright style which typified his writing he went on to savage government policy, both past and present, for selling short the domestic mining interest and continuing to deprive it of the resources necessary to enable it to respond to the current emergency. "We are now faced with the consequences of follies committed in the past 20 years, during which the greater part of our native base metal mining industry has disappeared. Were it not for that fact Cornwall could now be responding in greater style to the National effort...". The neglect and obstacles to the expansion of the industry remained and continued to grow. "The most important of these are the acute shortage of labour, the inadequacy of the controlled price the mines are receiving for their products, the difficulty of obtaining machinery and other supplies, and finally, taxation and capital restrictions". Examining each of these problems in turn, he concludes dismally that, "unless this attitude is revised...it is certain that additional production is not going to be forthcoming".

Nevertheless, the government had expressed a strong interest in propositions that could be speedily brought into production and Jack went on to review a range of prospects, from working and equipped but dormant mines to alluvials and dumps. His conclusion on prospects from the latter strongly reflected his views about the abilities of the "old men", set out in Section 2. They had looked carefully and worked with great efficiency. Notwithstanding the government's hopes and aspirations, he was sure that, "there can be little doubt that payable alluvials and dumps are now very definitely a thing of the past".

Jack concluded the first part of Section 2 with an unqualified denunciation of bureaucracy and geologists as obstacles to rapid and meaningful progression. "There seems to be a very real danger of endless delay and discussion by the appointment of committees and sub-committees

Introduction

and geological advisers to go over the ground for the hundredth time". In spite of the fact that some of the world's best economic geologists had been consulted on Cornish mining problems during the last 30 years, "the fact remains that professional geological advice during that period has failed to yield a single discovery of value...if the authorities need the metal which Cornwall can produce they will, in the author's opinion, be well advised to appoint men of action who are familiar with the County and its problems and give them power to act and make decisions without waste of time listening to theorists sitting in Committees". Clearly Jack was unabashed in job seeking when he thought his talents were needed!

Having established the general context of the increased demand for domestic minerals in the first article, Jack went on to look at his subject in more detail in a series of further publications. Section 2 Part 2 initiates this review with an outline of the past history of mineral production in Cornwall followed by a look at the geology of the southwest and the nature of its mineralisation. Noting the extreme antiquity of mining in the district, he looks briefly at its historical distribution and goes on to outline the changing balance of the principal metals in production, from copper to tin, during the third quarter of the nineteenth century. Unlike most current mining historians, his analysis continues uninterrupted through to the 1940s and serves to remind us that the twentieth century history of mining remains largely untold. We naturally find his comments on the reliability of the early official Mineral Statistics not to our taste, but then Jack cannot be expected to have always been right! In his comments on the geology of the region and the zonal sequence of ore deposition, he takes up and develops theories first sketched in outline in Section I. These two sections should certainly be read together. Although Jack has much more to say here than earlier, particularly in terms of examples of the zoning model, and also its limitations and exceptions, it is clear that he had not altered his basic views in any significant way. One important new but short inclusion related to lateral mineral development remote from the granite outcrops. Jack observed that many had argued that if a *downward* progressive through the killas to the granite often produced a changing progression of minerals, other lodes should exist emanating out from the *sides* of the granite into the killas at depth. "It has been widely argued that as the granite probably underlies the whole killas surface of the county it would only be necessary to sink sufficiently deep on the outskirts of the old estab-

Introduction

lished mining areas in order to intersect additional lodes in or close to the margin of the granite. It was hoped that vigorous development of such lodes in this zone would lead to the discovery of further important reserves of tin". While Jack probably found this concept attractive, he was obliged to admit that, "deep and costly developments at Tolgus, East Pool, South Crofty and Roskear have done much to discount the truth of this theory". To that time, "the available evidence is strongly suggestive that in Cornwall the major ore-bodies, and especially those yielding tin, are to be found relatively close to the granite outcrops, or where that rock is at no great depth below the surface of the land". While lateral *deposits* may not have offered great prospect for the future, Jack certainly thought that lateral *exploration* might. At the very end of this section, where he addressed the possibilities of funding entirely new mineralised areas, he concluded that "on the whole it seems much more preferable that most future discoveries will be confined to the established mining districts and that they will be obtained by deeper sinking in the shallow mines and by more extensive lateral development in or to intersect parallel ore bodies". In his only serious criticism of the methods of the "old men" he declares that the failure to use adequately cross cuts "to explore for new and additional lodes while exploiting known bodies has been one of the most serious and frequently repeated errors in the whole history of Cornish mining". Otherwise, the latter part of Section 2 Part 2 resembles the first part of Section 1 very closely and, at times, almost suggests the judicious use of scissors and paste on the earlier article!

Section 3, which is comprised of the remainder of the wartime articles, is easily the largest in this book. It is also the one that requires least introduction. Having outlined the background of his study in earlier sections, this one provides a detailed regional survey of the mining districts of west Cornwall. It is most unfortunate that the end of the war cut short his gradual eastward progression but at least he provides us with a view of those mines not covered in more recent surveys, such as T.A. Morrison's *Cornwall's Central Mines*. The sketch map on page 49 provides a guide to the discussions, which starts with a survey of the mines of the St Just and St Ives districts. It is notable that at points in this survey Jack looks again at mines mentioned in earlier articles but now has the opportunity to give them much more detailed and sustained consideration. Levant is one such working, first referred to in the latter

Introduction

part of Section 1. It was one of the very few exceptions to his rule that most old tin mines had been worked out by the "old men". There was clear evidence that its closure "was in no way due to impoverishment or exhaustion of the tin". In both sections he referred to it optimistically as a still promising prospect. Here he concluded that, "Almost alone among the abandoned mines of Cornwall, Levant may reasonably be regarded as more of an investment than a mining speculation...As a purely personal expression of opinion the writer most earnestly hopes that the neighbouring Geevor company will one day undertake this task". So keen was Jack on the Levant prospect that he again returned to the subject in the 1951 articles. At the beginning of Section 4 he re-examined this old but potentially still rich mine and concluded, that, "if the mine is ever worked again, at still greater depths, the best solution of the ventilation and ore-handling problems may well be the sinking of a major inclined shaft to give direct access from surface to the deep undersea parts of the mine". Geevor was to adopt a very similar solution for their own development at depth more than 20 years later. As far as Levant was concerned, however, the re-exploration work conducted there by Geevor in the 1970s and 1980s was rather disappointing, though it was discontinued as a result of the tin crisis before it was effectively completed. In his account of the earlier working of Levant and the reasons advanced for its failure to exploit fully the lodes, Jack gives us a glimpse of one of his other great prejudices viz. a dislike of the cost-book form of organisation. It was the adherence of the "old men" to this peculiar unlimited liability form of organisation that produces one of his very few criticisms of them. He declares that the nineteenth century working of Levant, "was a classical example of improvidence on the part of a "cost book"... company that had reaped enormous dividends from a trifling initial outlay without making proper provision for the future".

In his survey of the mines, Jack considered most of the principal workings of the districts, both promising and unpromising. In the far west, he was particularly enthusiastic about the future prospects for Levant, Geevor, Wheal Trenwith, Giew and Wheal Reeth, which he described as "amongst the best in Cornwall". For the Marazion district he took a very long look at the Rodney -Hampton-Tregurtha Downs group of mines, drawing extensively on the evidence of past managers and miners, to assess what he saw as the strategic core of the area. In the Breage district, Wheal Vor and its neighbour, Wheal Metal received considerable attention. Although undoubtedly excited by the prospects for further exploration at these once highly productive mines, Jack is more cautious than Collins in warning that there were "more shallow and easily proved prospects awaiting development elsewhere in the County". Still within that district, however, he was more encouraging about Godolphin, which he declared to be "one of the best speculations for tin that can be found in any part of the County", and Godolphin Hill area generally which, "has never yet been systematically explored, although it has long been thought to be an exceedingly promising speculation". The survey of the St Erth, Gwinear and Crowan district produced many mines, "that merit the expenditure of much new capital", and for the Wendron district Jack concluded that, "on the whole the prospects are not encouraging." Some merit was found for Binner Downs and Crenver and Abraham in the former district but only Porkellis and very possibly Polhigey were given any prospect in the latter. In his comments on all of the mines, both promising and unpromising, Jack provides useful material on the history of past working and often includes valuable data on the water problems experienced by the mines and the methods and costs of pumping. The latter may prove particularly interesting to those interested in the history of the Cornish beam engine.

Finally, in Section 4, Jack returns to a second regional review of the mining districts, this time not confined to the west of the County but embracing the most promising prospects from St Just to Callington. The first part of this review, prompted by "the growing pressure on base-metal supplies" from the late 1940s, covers the same area of western Cornwall considered in Section 3. It therefore clearly repeats much of the material presented earlier for mines such as Levant, Giew, Wheal Trenwith, Godolphin Hill, Wheal Vor and Porkellis but also introduces interesting new information on the prospects for Wheal Racer. As he extended gradually eastwards from the boundaries of his earlier study, Wheal Grenville and South Condurrow were included, as well as the eastern part of East Pool and Agar. These properties were later taken over by South Crofty but their redevelopment was interrupted by the tin price collapse of 1984. The Scorrier area, largely drained by the Great County Adit, was described as "one of the best, if, indeed not the most promising large mining area to be found in the whole of Cornwall". While just to the north, the Stencoose district was said to be "still practically virginal and seems well worth thorough investigation".

Introduction

On the north coast, Wheal Coates, Wheal Kitty, West Wheal Kitty and the Perranporth mines also received detailed attention and optimistic praise and the Great Perran Iron Lode was again singled out for investigation at depth. Although he frequently disagreed with Collins on other sites, Jack reminded his readers that, "the possibilities latent in Collins' suggestion are so great that the matter ought to be tested by a comprehensive series of drill-holes put down to intersect the lode at depths of at least 1,500 ft. Even if the lode failed to yield copper in depth its great possibilities for blende should not be overlooked in view of the increasing scarcity of zinc". After looking at further prospects in the Newlyn East, Mithian and Perranzabuloe districts, Jack concluded with a look at mines of east Cornwall. Here Holmbush, Kelly Bray and Redmoor were recommended for further development at depth, the hope of finding more payable copper and lead at depth as well as the good prospect of increasing tin values. The Prince of Wales mine received particular attention as a good prospect for silver ore as well as tin and was strongly recommended for "far more extensive trial than it has yet received". Nearby Trebartha-Lemarne was also recommended but probably with less real prospect than the other mines of the district.

In putting this collection together we have attempted to give it book format rather than to present it simply as a collection of articles. Certainly there is some repetition of material but it comes together well in this new form. All of the maps, tables and graphs presented in the original material are reproduced here together with a selection of the photographs. We are grateful to the *Mining Journal* and Mr J. Bullen for permission to reproduce their copyright material.

SECTION 1: CORNWALL'S UNDEVELOPED MINERAL RESOURCES IN THE 1930s

For the purpose of this article, which is an examination of Cornwall's mineral resources, only the metallic minerals are considered as such, thus excluding "earthy" substances like China Clay and various types of stone which, in the popular sense of the word, are not "minerals". An immense number of ores have been discovered and worked on a commercial scale in Cornwall, but those of tin and copper predominate. Arsenical and tungsten ores, the latter especially during the present century, have become increasingly important by-products of tin and copper production though rarely found in payable quantities by themselves. Following the primary ores those of lead, iron, zinc, manganese and uranium are the most valuable commercially, and in that order of importance. Silver can be correctly regarded as a by-product of lead, and to a lesser extent of copper and zinc rather than as a separate group of silver minerals in which form it is comparatively rare. Of the remaining metallic ores which have been discovered in small quantities none have been of great importance, and in a general review of the subject they do not merit further consideration especially as there are no indications of their existence in important quantities.

As the geology of Cornwall has a direct bearing on the position and nature of its metallic ores, it is necessary to review the main geological features before considering the future possibilities of mining for the most important metals as enumerated above.

Basically Cornwall consists of sedimentary rocks, which may once have been fine silts or mud deposited at the bottom of an ancient sea until, under pressure, they became solid rock. At a later geological age eruptive ("igneous") rocks, the most important of which is granite, made their appearance from below. The granite, quite apart from local intrusions which now form the greater part of the high moorland and hills of the County, is probably present everywhere at no very great depth, and forms the foundation on which the sedimentary rocks, or "killas" (as the miner terms them collectively), now rest. At the time of their intrusion these igneous rocks were probably at exceedingly high temperatures or

Section 1: Cornwall in the 1930s

even molten and their effect on the overlying or surrounding sedimentary rocks was so intense that great chemical and physical changes took place in the latter. The rocks thus altered or metamorphosised now form a halo-like band or metamorphic aureole of varying width around the protruding granite masses and above the granite where it is comparatively shallow though still hidden by over-lying killas. The subsequent solidification and cooling with consequent contraction of the intrusive rocks set up such intense strains in the Earth's crust that great fissures and other earth movements took place, thus forming in the intrusive and overlying killas rock fissures and channels which later became charged with metallic minerals thus giving rise to the economically valuable deposits of mineral of the present day.

As might naturally be expected, the greatest area of stress and fissuring occurred at or close to the margin of the granite intrusions, in other words, in the metamorphic aureole previously mentioned and in the outer "skin" of the granite itself, and it is just in or close to this granite-killas junction that the overwhelming bulk of the economically important minerals occur. Valuable deposits of mineral in the killas far from the nearest granite outcrop and where granite is probably at relatively great depths below are as rare as are important deposits 2,000 feet and more below the original surface of the granite. In general terms it is noticeable that the metallic minerals deposited in the veins are in a definite sequence or zonal arrangement. Commencing with tin oxide (the only commercially important ore of tin) which occurs in or close to the granite, as the margin of the latter is approached tungsten ores make their appearance, and then those of arsenic, followed by ores of copper in the killas with tin tending to decrease with increasing distance from the granite. Finally the copper tends to die out and is succeeded upwards in order by ores of zinc, lead, iron and manganese. There is little doubt that many of the famous veins or lodes once contained many or all of these zones, but during the passage of the geological ages the upper portions of the land have gradually been worn away so that the zone of a lode which is now exposed at surface may be the copper part or, as in the case of the lodes of the Wendron District, the underlying granite is now exposed and even much of that has been weathered away with the result that only the "roots" or deepest portions of the lodes are now intact. The economically important factor in the Wendron area is that the depth of the mines is thus severely limited, and in practice they

Section 1: Cornwall in the 1930s

have mostly proved to be very shallow deposits.

At one period, certain geologists were inclined to argue that as the granite probably underlies the whole of the killas surface of the County it was only necessary to sink sufficiently deep in order to get close enough to the granite margin at any given place, no matter how remote from its nearest outcrop at surface, in order to discover further mineral bearing veins, with the ores in them arranged according to the afore-mentioned zonal sequence. Unfortunately, deep and costly developments carried out in the Camborne-Redruth District in recent years have failed to prove this theory, and there are now good grounds for thinking that the bulk of the economically valuable ores in Cornwall lie relatively close to the margin of the granite bosses as at present exposed by "denudation" or weathering of the surface. If, as now appears, this theory is generally applicable to the ores of tin and copper, that it is within those districts where the killas-granite junction occurs at a lesser depth than 2,500 feet (and especially where the nearest granite outcrop is not more than a mile distant) that we may expect to find the more profitable deposits of tin and copper. Incidentally it is worth noting that the zonal theory of the sequence of ore deposition is not everywhere applicable in Cornwall, as some of the greatest deposits of copper have been found in granite and the killas has yielded some of the best tin lodes.

The area of the County within which it is advisable to seek for metallic ores having been partially defined, it is necessary to note a further natural and hitherto unexplained fact which has to be taken into account, viz. that of the total length of the margin of the granite intrusions in Cornwall, only a very small percentage is mineralised. There are many miles of "granite margin" where there is an almost complete absence of metalliferous veins of any sort whatsoever. It would, therefore, only be spending money and time uselessly to develop at depth such areas and, in the main, the search for unworked deposits of ore is brought back to the old and existing mining districts in or close to the margin of the granite intrusions.

Two further types of area also offer some possibilities of development viz. "granite" districts which, as a result of denudation, are now far remote from the once overlying killas and "killas" districts far from the granite outcrops. The former once produced a few important tin mines with many lesser ones, the latter have been important producers of copper and more especially of lead-zinc-iron veins, but the future pro-

Section 1: Cornwall in the 1930s

ductivity of both types of area is now doubtful. In the first case, where tin veins occur in the granite hills, it is only the roots or lowest portions of the veins which now remain and as the upper parts have been weathered away it follows that all such veins outcrop at the present surface and in most instances they have long ago been worked out by previous generations, especially as lodes outcropping at surface were far more easily detected in the past than in later days when the land had been extensively cultivated. Much the same may be said of the improbability of discovering previously unworked deposits in the killas far remote from the granite. Such metalliferous veins have been comparatively rare though they have produced many of the foremost lead deposits. The latter have often been found in the "cross-courses" (extensive fissures running through the country at approximately right angles to the bearing of the tin and copper lodes in the nearest granite area) which are usually barren in metallic ores close to the granite and far remote from the latter rock, yield only lead, zinc, copper or iron but never tin. Their chief ore content, lead, usually dies out at the relatively shallow depths of 800 to 1,000 feet and, as most of such known deposits have been extensively explored from surface to the greatest depth at which it was found this metal existed in payable quantities, there now seems little opportunity of fresh discoveries in these types of lodes.

A highly mineralised and intensely disturbed and fissured tract of the Earth's crust, such as Cornwall, is bound to contain frequent surprises and exceptions to any rule which can be formulated, but, with certain notable exceptions it would appear that any future discoveries of metalliferous ores, more especially those of tin and copper, lie within the old and well recognised mining districts at or close to the margin of the granite intrusions or where the surface of the latter is at no great depth from the present surface of the land. If the probability of new ore discoveries in Cornwall lies within these old mining districts, the obvious question which arises, especially in view of the many failures which have resulted from the re-opening of old mines during the 20th century, is "whether there is any further prospect in these areas, particularly in regards to tin, and whether previous generations have not completely exhausted them?" The answer to that question can only be obtained by a further examination of the available facts.

When discussing the zonal sequence of the various ores in the lodes it was seen that tin usually occupies the lowest horizon, which is generally

Section 1: Cornwall in the 1930s

in or close to the granite, and that where there is killas superimposed on the granite the upper part of the vein in that rock is more usually productive of copper ores. In a very approximate way the copper and tin mines of Cornwall may be divided into three groups:-

1. The copper mines which have worked the shallow upper parts of the lodes, mostly in killas.

2. The tin mines which have developed the lower parts of the lodes mostly existing in granite, now exposed at surface through long ages of weathering.

3. Copper and tin mines where following the exhaustion of the copper, at relatively shallow depths, tin was discovered by further sinking usually into or close to the granite rock below.

As far as the last two groups are concerned, unless there is good reason for thinking to the contrary, all those seeking fresh mineral deposits would be well advised to leave them severely alone. It is through re-opening so many mines of these types that the bulk of the failures in the County have been brought about in recent years. There are several instances where reckless finance, short-sightedness in mining, catastrophic drop in price of metals, or disaster in some form has closed down a mine which still merits re-opening and further development. But in the majority of cases the fact must be plainly faced that where the previous generations or "old men" worked a mine on a considerable scale for tin, either alone or subsequent to the shallower copper deposits, there is little prospect of making further payable discoveries, for the simple reason that the old men, like the present generation, continued to work and explore a piece of ground as long as they thought there was the slightest prospect of it being or becoming profitable. They were just as astute, and far more capable, than present-day mining men in the art of working successfully relatively small and very complicated mineral deposits such as constitute the majority of the Cornish mines. And, during the 19th century, when the greatest mining activity prevailed in Cornwall, though the price of tin was on the whole very much lower than that ruling in the 20th century, still, wages and all costs were proportionately low, and former generations could frequently prosper just as much with tin at £100 per ton as present-day companies can do with the price considerably over £200 per ton.

Furthermore, it is an oft repeated fallacy that the "old men" could not concentrate the ores as successfully or cheaply as can be done today,

Section 1: Cornwall in the 1930s

whereas, with the exception of the invention of the magnetic separator which solved the difficult problem of separating "Wolfram," the chief ore of tungsten, from tin in the relatively few mines which contained wolfram in quantities, we are probably no better and no worse than the old men as far as concentration is concerned. What we now do with machinery they did by hand, and with very cheap labour at that. Therefore, as stated before, unless a detailed knowledge of the history of a mine provides good and convincing evidence as to why it should be re-opened it is far better left severely alone if in the past it has been a large producer of tin and worked principally for that metal.

However, as far as future discoveries of tin are concerned, the best prospect, and indeed a very promising one, seems to lie in the re-opening and deeper sinking of the first group of mines — namely, the relatively shallow copper mines in killas rock which were closed down on the exhaustion of the copper at a time when it was not generally realised that tin was frequently to be found in even greater quantities below the copper. But in making that statement a very strong qualification must be added, viz. that only those copper mines should be re-opened which are in killas and which appear to be relatively near to the granite in depth, or are in districts renowned for tin as well as copper production. In the existing state of knowledge on the subject it can only be said that tin does not always follow copper in depth, there are considerable areas in the old Parishes of St Erth, Phillack and Gwinear, and in western Devon, which geologically is a part of Cornwall, where there have been immense deposits of copper worked to fairly considerable depths without, however, finding granite in depth or any trace of tin in commercial quantities underneath the copper.

In giving actual examples of districts where copper has been the chief product down to the very moderate depths attained in killas, and where granite probably lies not far below, one of the first to deserve attention is the Scorrier mining area east of Redruth together with parts of the adjacent Parish of Gwennap. The Scorrier District consists, from a mining and geological point of view, of all those mines from Wheal Harmony and Treleigh Wood, just north east of Redruth, to Wheal Busy between Scorrier and Chacewater. In addition to those mentioned the District contains amongst many others such mines as Great North Downs, Wheal Rose, Wheal Briggan, Hallenbeagle, Old and North Treskerby, Wheal Chance, Wheal Prussia and Cardrew and the Peevors, the latter

Section 1: Cornwall in the 1930s

being at one time notable tin producers. Most of the mines in the area have been worked solely or principally for copper, they are all moderately shallow, 1,000 feet or less from surface, and the majority have only been worked in killas rock. The Carn Marth granite boss outcrops at surface on the southern margin of the area and is probably at no great depth below the bottom of most of the mines under consideration. Furthermore, quite apart from discoveries of tin, it is recorded that many of the lodes were showing considerable quantities of arsenical and tungsten ores in the bottom workings in several of the mines named, and this is a most favourable indication of a change from the copper to the tin zones in depth. In fact the area strikes many observers as being one of the most promising pieces of ground now remaining in Cornwall for successful large scale mining.

Transferring the attention to West Cornwall, the next extensive area to be mentioned is that which stretches from Marazion eastwards to Prah Sands and from the coast inland for 2 or 3 miles. The latter district is somewhat heavily watered in places, which increases working costs, but the numerous mines are mostly shallow and there would appear to be some very promising prospects in this neighbourhood. Not very far north and east of the last tract of ground a favourable area would also appear to be that north of Godolphin Hill containing the old Godolphin mine and eastern part of the West Godolphin mine together with other properties around Leedstown still a little further to the north. In this connection special mention should be made of the Binner Downs and Crenver and Abraham mines. From the known outcrops of granite it would appear probable that this rock lies at no great depth below any of the mines mentioned extending from Marazion to Leedstown, while those around the Godolphin Hill are actually in or very close to the granite of that hill.

While still in the western part of the County it is worth noting that on the north coast almost in the town of St Ives itself there is the old Trenwith copper and uranium mine which merits a further trial in depth, and possibly even more so the piece of ground westwards between Trenwith and the St Ives Consols mine. This section of the lodes extending from Trenwith to the Consols has never been developed in depth and contains the killas-granite junction throughout its length.

Further east, on the north coast, there are the extensive old copper mines in and west of Perranporth. In view of the presence of the

Section 1: Cornwall in the 1930s

granite outcrop at Cligga Head and the fact that granite was found in mining operations not far below sea level a quarter of a mile to the east of Cligga there is good reason to expect these shallow copper mines hitherto worked in killas, to come down on granite at no greatly increased depth with a promising prospect of changing into tin producers. A small quantity of tin has already been sold from these particular mines.

Transferring the attention to the opposite coast, a most favourable tract of mining land is to be found between St Austell and Par extending for about 2 miles west of the latter place, and from the coast inland for approximately 1 mile. Throughout the greater part of its length this district contains prolific old copper mines under which the granite probably lies at no great depth though it declines towards the east. The mines at the western end of this group have already been very successfully worked for tin in the past.

Finally, in the eastern part of the County there are several mines on or close to the northern and southern slopes of the celebrated Kit Hill-Hingston Downs granite mass which would appear to merit further attention, not forgetting the group of mines at Kelly Bray on the western end of Kit Hill itself. Several of the mines in this area have been big tin producers in the past and there are others which appear to be only just getting down into the tin zone which well deserve a further vigorous trial.

However, even when one has exhausted the list of shallow copper mining districts which are probably destined to change from copper to tin bearing and from killas to granite in depth there still remain notable exceptions which, though they do not come under this general heading, are yet still worthy of further trial as they were only shut down because of some peculiarly local or individual reason. The more important of these mines and districts deserve passing notice.

Commencing in the west, the first to attract attention is the famous Levant mine which extends under the sea near Pendeen Lighthouse for about $1\frac{1}{2}$ miles. Here the great courses of tin and copper ore run in a north westerly direction on and about the killas-granite junction, the principal ore bodies being in the killas. Its peculiar situation under the ocean made it an expensive mine to operate because in the past no long-sighted provision had been made for the future, with the result that the ore, coming from a mine ever deepening and extending further and further under the sea had to be rehandled so many times before it reached

Section 1: Cornwall in the 1930s

the surface that it became hopelessly uneconomic whenever tin slumped badly in price. The necessity of facing the large capital expenditure required for sinking a new main shaft direct to the bottom of the mine was repeatedly shelved by the company, especially during the last 25 years of its existence. As a result of this lack of foresight the company collapsed and the mine was abandoned in September 1930. The end was hastened by the abnormal wartime conditions, which nearly crippled mining in Cornwall, followed by a terrible disaster in 1919 with heavy loss of life, together with the post-war slump which quickly followed, succeeded by years of financial stringency and, finally, the disastrous slump in tin which in 1930 reached about the lowest price for the century. As far as can be judged, Levant has yet a long and profitable future before it if and when the capital can be found to re-open and work cheaply the deeper parts of the mine. It is certain that the closing was in no way due to impoverishment or exhaustion of the tin. The average grade of ore is unusually high, considerably above the average of any mine worked in the County for many years past, and the water to be pumped is singularly little. In addition to tin the mine is still likely to produce an appreciable quantity of copper and arsenic, and it may reasonably be said of Levant that few mines in Cornwall better deserve the cost of re-opening.

The bulk of the tin mined in Cornwall has come from lodes within or close to the granite, but there are two areas which are very notable exceptions to this general rule and where tin has been mined in very great quantities and richness from lodes in killas. These important exceptions are found at St Agnes on the north coast and in the great killas mining area immediately east of the Tregonning-Godolphin granite hills in south west Cornwall. In each area there are still very promising prospects worthy of further investigation. At St Agnes, though there is otherwise little of promise, the great "shoot" of tin under the valley between Wheal and West Kitty is still deserving of a vigorous re-working. Operations here only ceased with a very fine lode of tin in view by reason of the disastrous slump in the price of tin in 1930. Extending from the western slopes of St Agnes Beacon to the cliffs just east of Chapel Porth is the isolated tin mine known as Wheal Coates. The granite core of the Beacon is probably here at no great depth but up to the present a large medium grade and cheaply worked lode has been mined to shallow depths only in killas and an "Elvan" dyke, the latter being an igneous intrusive rock closely allied to granite. Though worked or re-opened at several

Section 1: Cornwall in the 1930s

periods, operations here have never been conducted with much vigour or financial backing, and the property seems well worthy of a further and more energetic trial.

Returning to the great killas mining area which lies to the east of the Tregonning and Godolphin Hills, near Helston, it is worthy of notice that this contains such outstanding mines as Wheal Vor and Wheal Metal, more properly known as Poldown. Wheal Vor, which is an exceedingly ancient mine, was re-opened in 1812 and subsequently became one of the greatest tin producers ever known in Cornwall. The monthly production at one time reached the extraordinary total of 220 tons, over 1,100 people being employed. The low prices for tin then prevailing and legal troubles brought the enterprise to a close in 1847, but the mine together with Wheal Metal and a great many other properties in the district was re-opened in about 1852 by the Great Wheal Vor United Mining Company. This grandiose and ill-advised concern worked all the mines under its direction with such utter recklessness and lack of judgment that all had been abandoned within a few years with the exception of Wheal Metal which had turned out to be extraordinarily rich in depth. However, the mine was worked without the slightest regard for the future and was finally closed in 1874. A detailed examination of the recorded features of the mine together with an appreciation of the improvident manner in which the property was operated is strongly suggestive of excellent prospects yet existing here, especially in the direction of parallel lodes to those which the late company worked. Very little cross-cutting and lateral development seems to have been carried out and the mine was only worked as long as it would pay its way with the minimum of development, after which it was promptly abandoned. The same may be said of Wheal Vor and the intervening ground between the two mines. The unsuccessful attempt to unwater Wheal Vor about 30 years ago only reflects on the machinery and methods employed at that time. The water was only partially pumped out and the prospects remain as unproved as if nothing whatsoever had been done.

It remains to be pointed out that there is at least one most promising tin mine in West Cornwall which is in granite rock from surface and where there still appears to be excellent prospects. The property in question is about 2 miles S.S.W. of St Ives and it has been variously known as South Providence, Reeth Consols, or as the Giew mine, under which title it was last worked, operations coming to an end during the

Section 1: Cornwall in the 1930s

post-war depression through the low price of tin and the very high cost of all materials and supplies at the time. Only one of two lodes has been extensively worked, and all the available evidence is suggestive that the lodes are likely to continue productive and profitable for tin further to the east on the eastern side of Trink Hill. The bottom workings extend to nearly 1,500 feet from surface and the future prospects seem to lie more in their lateral extension eastwards than in depth. This piece of ground has long been regarded by mining men as an excellent area and the last re-working went a long way towards corroborating that opinion. If it had been possible at the time to find the capital necessary for sinking a new shaft on the other side of the hill it is probable that the mine would have long continued in successful operation.

Before leaving the district in which Giew is situated, it is worthy of notice that there exists, almost at the summit of the granite mass of Rosewall Hill, about $1\frac{1}{3}$ miles north-west of Giew, a small and shallow mine known as Wheal Racer which is thought locally to be worthy of further trial. The mine which is shallow and has very little water, contains two lodes which are converging in depth, a condition often favourable for the enrichment or enlargement of the lodes. The latter are of fair grade, fairly narrow as is general in this district, but cheap to mine, and it would seem that a further trial of this small mine is warranted.

In concluding a review of those mining districts and certain individual mines which seem well worthy of further development, there remains to be mentioned what, as far as dimensions are concerned, is probably Cornwall's greatest lode, namely, "The Great Perran Iron Lode." This exceedingly large vein was referred to by writers on mining and geological subjects as long ago as 1758 and, though for a variety of reasons it has never been proved for more than about 420 feet from surface or 180 feet below sea level, yet its future potentialities as a great copper and tin lode in depth are very great. The lode outcrops in the cliffs at the north end of Perranporth Bay and strikes across the County, first in a direction of 30° south of east, for 2 miles, and then bends around until it runs in a nearly east to west direction. It has been seriously worked in a number of mines for 3 miles from the coast and trials on it have been made for at least another $1\frac{1}{2}$ miles eastward. Beyond that point its course is somewhat uncertain, though it is said that it has been traced as far as the Parish of Grampound, a distance of 12 miles altogether. It dips to the south at an average angle of about 50° and the width, though

Section 1: Cornwall in the 1930s

unusually great, is exceedingly variable, usually anything from 3 to 20 feet and occasionally as much as 40 or 50 feet or even more. Throughout its whole course the lode is in killas except where it is crossed by an elvan and though the nearest granite outcrop at Cligga Head, west of Perranporth is $2\frac{1}{2}$ miles distant from the lode, it may be that the granite underlies the whole of the district at no great depth.

The majority of the great copper lodes in Cornwall, many of which afterwards became the most prolific tin producers in depth, had at or near surface a very similar composition and appearance to this great lode, and it has long been the opinion of many eminent mining men that when it is explored in depth it will prove to be far more of a copper and tin lode than its present-day name suggests. The "back" or upper portion of a lode becomes very much altered and chemically changed when exposed to atmospheric influences and water charged with carbonic acid and other gasses, absorbed from the atmosphere, seeping down through it to the drainage or water level of the surrounding country. The bulk of the copper and certain other lode contents are dissolved, carried down, and re-deposited at greater depths in the vein. The chief remaining metalli. contents of this upper part of the lode then consists of altered iron compounds and this type of yellow or brown cindery looking lode is termed a "gozzan". Most mining men familiar with the appearance of the great "gozzan backs" of other productive copper lodes in Cornwall have been of the opinion that the Perran Iron Lode was in reality only a great "gozzan" indicator of an exceedingly important copper and tin lode at greater depths, rather than a true iron lode in itself.

The greatest production from the 7 or 8 shallow mines and open-cast workings on this lode was obtained during the years 1854–1894, especially for iron ore when the latter was sufficiently valuable to render the working of the lode for that material a commercial proposition. Since then the lower prevailing prices for iron have rendered it unprofitable on the whole, and its size and the necessity of working it on a large scale seem to have discouraged anybody from making a serious attempt to develop it in depth. Nevertheless, the possibilities of success are very great, and in these days when diamond drill holes of great depth can be quickly bored it would seem to be an excellent speculation if a syndicate could be formed to put down several holes as a preparatory step, intersecting the lode at, say, 900 to 1,200 feet from the surface. It is quite possible that Cornwall's premier tin and copper lode is just awaiting somebody

Section 1: Cornwall in the 1930s

with sufficient vision and means to give it a vigorous trial in depth. It is a matter of historical fact that some of what afterwards proved to be the best lodes in Cornwall remained untouched until well into the 19th century, though their "gozzan" backs were visible to all and sundry, and the present neglect of the Perran Lode may be a case of history repeating itself.

In passing, it may be noticed that in 1910 it was estimated that this great lode had yielded since 1854 at least 200,000 tons of iron ore, 32,000 tons of zinc ore, 2,500 tons of lead ore, and 250 tons of copper ore, in addition to small quantities of ores of other metals. It is worthy of note that even tin has been seen in the lode, and it is strongly suggestive of the sequence of metallic ores deposited according to the afore-mentioned zonal theory.

It will be noticed that this examination of Cornwall's mineral resources has so far been concentrated on an attempt to discover where profitable deposits of tin may be found. The natural question therefore arises "what are the prospects of successful discoveries of the other metals for which Cornwall has been so famous?"

As previously emphasized, arsenical and tungsten ores are, almost without exception, merely by-products of tin and copper ores and as such they are only profitable when mining is being conducted for tin or copper. It is fairly certain, however, that large quantities of each ore are still in existence in the County, but any attempt to estimate their quantity would be futile. The amount of each which will be produced in the future will probably depend almost solely upon the activity of tin and copper mining, especially the former. It must also be borne in mind that not all the lodes in the County contain arsenic in any quantity and tungsten ores are even more scarce, being confined to a relatively small number of mines. If it were desired to discover the ores of tungsten in the maximum quantity available in any one district the Scorrier area would probably be the most promising one to develop for that purpose. The mines on the eastern border of Cornwall are probably richer in arsenical contents than those of any other area in the County, the Redruth-Camborne district not excepted.

As regards copper, lead and zinc, though the first mentioned is to an overwhelming extent the most important of the three, all are equally well considered together as far as future prospects are concerned. Quite apart from any question of price, which for the last generation or so has

Section 1: Cornwall in the 1930s

settled down and looks likely to remain for a long period at a level too low to make the mining of the metals profitable in Cornwall, there remains the doubt whether, with the possible exception of the Great Perran Iron Lode, there is the likelihood of any important discoveries of any of these metals being made in the future. As pointed out, when the zonal theory was under discussion, these minerals are generally found in the upper parts of the lodes, though probably much of the lead production has come from the "cross-courses" or veins at right angles to the tin and copper lodes. In any case nearly all the known deposits of copper, lead and zinc have been comparatively shallow and the lodes containing them were mostly exposed and visible at surface, and consequently received far more attention and development in the old days than many of the deeper tin lodes, which have sometimes only been discovered by lateral exploration from other previously worked deep lodes. Consequently, with the possible exception of the Great Perran Lode, there is the strongest reason for doubting the probability of any further important discoveries of either of these three minerals being made in the future. There is always the uncertain chance of new discoveries, but the tremendous fall and virtual cessation of copper production in Cornwall between 1856 and 1887 is very significant. It has to be admitted that the price was falling during these years of decline in production, but the sharpest fall in output occurred while the price was still high, especially as judged by 20th century standards, and this in itself is an eloquent proof of the rapid exhaustion of the copper that was taking place at the time and the improbability of making important discoveries of copper in the future. That, however, does not exclude the possibility of working the remains of the old copper deposits as a by-product in the future when many old mines are being operated at greater depths for tin.

As regards iron, most authorities are of the opinion that many of the Cornish deposits, as in the case of the Perran Iron Lode, are more in the nature of Gozzans than true iron deposits. Furthermore, the relatively low price of iron ore prevents the working of it being commercially successful unless it is of suitable quality and purity and can be mined in very large quantities. It is doubtful if any of the Cornish deposits possess these qualifications to a sufficient degree with the possible exceptions of the Perran Lode, the Restormal Iron Lode near Lostwithiel and the very rich red haematite lode which has been worked at the Ruby Iron Mine and elsewhere in the decomposed china clay granite north and east of

Section 1: Cornwall in the 1930s

St Austell. In each of the lodes mentioned a great tonnage of iron still probably exists at comparatively shallow depths.

The possibility of discovering manganese in quantity, judged by past results, is small. Where found it is mostly associated with the iron ores of the County.

From old records it is apparent that a very considerable tonnage of uranium bearing ores must have been raised in the past, especially from certain of the copper mines, but at that time, having no commercial value and being merely an impurity in the copper ores, it was rejected and thrown aside whenever possible. Wherever extensive heaps of such ores were to be found on surface they have been picked over and treated during the present century. In addition, two or three old mines have been re-opened and worked on a small scale for the remnants of such ores left by the old men, but none of these ventures seem to have been a success and as far as the relatively rare uranium ores of the County are concerned it is probably a case of "bolting the stable door after the horse has gone."

Silver has never yet been found as a true silver ore except in very small though spectacular "bunches". The output from Cornwall came virtually from the lead lodes, and as such the future of silver production in the County is very problematical indeed.

The mode of occurrence and situation of the chief metallic minerals of the County of Cornwall have now been reviewed at length, especially with a view to forming an opinion of the probability or otherwise of any of these minerals still existing in important quantities. The examination of the evidence in the light of the existing state of knowledge of the subject suggests that tin, with its associated arsenical and tungsten ores, is the only metal which is likely still to exist over wide areas and in really large quantities, and so situated as to be mined successfully in the future.

The areas at present held and being mined and developed by active mining companies have been deliberately excluded, as they do not come within the category of undeveloped mineral resources of the County in the meaning of this review.

In conclusion, it would appear that the matter may be summed up by stating that the possibilities of making successful discoveries of tin in the County, more especially under the worked out copper deposits in certain selected areas, are very promising indeed. These areas, together with certain other classes of deposit where copper is altogether absent or

Section 1: Cornwall in the 1930s

as yet unworked still offer very favourable opportunities for commercially successful mining in the County of Cornwall.

SECTION 2: CORNISH MINING IN IN THE SECOND WORLD WAR

Section 2: Part I

THE INDUSTRY AT THE BEGINNING OF THE WAR

The spread of the war to the Far East, with the resulting embarrassment of the tin and tungsten supplies of the Allies, has been responsible for a revival of interest in Cornwall as a producer of these two vital metals. The authorities, which have apparently ignored the potentialities and war-time importance of domestic producers of these metals since the last war, are now openly interested. Many people, however, are frankly sceptical of the future of Cornwall as an important mineral producer, while the probability of being able to work such mineral deposits profitably within the County in times of peace is even more widely doubted. The prejudice against Cornish mining is hardly surprising in view of a disastrous series of unsuccessful enterprises opened up during the past 40 years which have largely overshadowed the few successes obtained during the same period. It has long been the author's view, however, that the greater number of past failures and the resulting disfavour into which the County has fallen are largely due to lack of knowledge of the Cornish mineral deposits and he is unable to share in the prevalent pessimism concerning the future successful working of a large proportion of these numerous ore-bodies. The present review is being attempted, therefore, with a view to emphasizing some facts concerning the Cornish mines and

Section 2: Mining and the War

ore deposits, to making known the most favourable locations in which new ore-bodies might be sought, and to formulating suggestions for the better working of the Duchy's mineral riches in the future.

Present Position

Before proceeding with a more general review it may be as well to deal at some length with the consequences of the Far Eastern War inasmuch as they immediately concern Cornwall. In this connexion it is obvious that anything that is written is likely to be completely out of date by the time it appears in print, but there is one fact that is certain to remain constant for a considerable while — namely, the imperative need to obtain from Cornwall the maximum amount of tin, tungsten, arsenic, and other vital metals that can be produced. The question becomes primarily one of supply and time rather than of economics — in other words, what Cornwall can produce within a limited period.

We are now faced with the consequences of follies committed in the past 20 years, during which the greater part of our native base-metal mining industry has disappeared. Were it not for that fact Cornwall could now be responding in great style to the National effort and making a really important contribution to our annual needs of tin, tungsten, and arsenic. Now, however, it is quite obvious that any increase that can be achieved in time to assist in the winning of the war can only amount to a very small percentage of the actual demand.

Before dealing with the technical aspects of increasing output it is essential that consideration first be given to obstacles which are preventing an increase from the existing producers and which would make themselves felt even more acutely were an attempt to be made to bring additional mines into production. The most important of these are the acute shortage of labour, the inadequacy of the controlled price the mines are receiving for their products, the difficulty of obtaining machinery and other supplies, and, finally, taxation and capital restrictions.

Section 2: Mining and the War

Labour

It cannot be emphasized too strongly that an increase of production is not going to be obtained by committee meetings in London or even in Cornwall. What is needed, and needed immediately, is an adequate supply of suitable labour. Everyone conversant with the situation realizes that the existing mines are critically short of men and that the major producers are living on their rapidly diminishing ore reserves. Under present circumstances the mining companies are powerless to help themselves in this respect and labour must be provided or the country will have to do without metal it urgently needs. The mining companies can be relied upon to do their utmost in the National interest, but they cannot achieve the physically impossible and those responsible for deciding upon the best allocation of existing man-power must now make their own decision.

It should be realized that hard-rock mining is a skilled job requiring men of a hardy and active type and that it is an occupation quite unsuitable for the employment of completely inexperienced and casual labour. Unlike many industries it cannot absorb and train in a few months a large labour force composed of men and women recruited from every walk in life. It is essential that the manner in which the authorities have handled this problem to date should be completely and immediately changed if an increase of output is to be obtained. The most certain means of achieving this end is to release from the Armed Forces the considerable number of men called up before the tin-mining industry was classified as an entirely reserved occupation. In the original schedule of reserved occupations miners under 23 years of age were not reserved and, in consequence, the mines lost the greater part of their younger men who, having just completed their apprenticeship as "boys", were then attaining their maximum value as miners. The actual number of men involved is now rather difficult to assess, but although they would be of the utmost value to the country as metal miners at the present time their actual numbers are so small that their withdrawal from the Forces would be of completely insignificant military importance.

The official view of the matter, however, is that if an exception is made for the tin mines every industry in the country will clamour for the return of its men and that the request cannot therefore be granted.

Section 2: Mining and the War

It is therefore obvious the authorities have not yet realized that notwithstanding the amount spent on the military training of a man who has inadvisably been called up he may be worth infinitely more to his country in his civilian occupation than while serving in the Forces.

At the present time the principal mines have long application lists of men who are prepared to return to their jobs if their former employers can obtain their release, but although an immense amount of time and effort has been expended in this direction the total number of men so far released is a mere handful.

Quite apart from the question of obtaining additional labour for the industry it is essential that men should no longer be actively encouraged to leave for higher-paid foreign mining jobs or for other Government work within the country. It was with amazement that the operating companies discovered that Cornish miners were still being encouraged to travel to African copper mines, even after the Japanese invasion of Malaya had commenced.

Effect of Metal Prices

The prices the mines receive for their products are now completely out of proportion to the increased cost of their supplies and they are consequently unable to offer sufficiently attractive wages to prevent a steady loss of men to far more highly-paid munition work. To argue that a man is prohibited from leaving a "Protected Industry" without a good and sufficient cause is totally beside the point. If a man is determined to leave it frequently happens that he succeeds in obtaining permission to do so by slacking until he is of no use whatsoever to his employer, who, being unable to compel him to work, is virtually forced to release him. If the mines were given a reasonable and more proportionate price for their output many of these labour difficulties could at once be solved. It also seems that the authorities would be well advised to abandon their insistence on the payment of a guaranteed high minimum wage in those mines where they are responsible for supplying additional labour. Human nature being what it is, contract work is indispensable for most underground operations and any attempt to fix a minimum inevitably leads to a disastrous falling off of effort and output. At two mines, where

Section 2: Mining and the War

a few men were provided on condition that this minimum was guaranteed, the principle led to much illfeeling amongst the men already at work and, as might have been foreseen, quickly led to a general falling off of output, which in one case at least contributed to the eventual closing of the mine.

Apart from the acute labour shortage the most serious hindrance to any immediate increase in production is, as previously mentioned, the low price the producers are receiving for their products. Generally speaking it can be said that the price which the mines are receiving for their tin and wolfram output is little better than that prevailing immediately before the war, while the price of arsenic is so low that it does not pay to mine. Were it not an undesirable byproduct that must be eliminated from the other concentrates there would be none offered for sale at all.

As an example of the disastrous position in which the companies are now placed the following facts concerning one of the largest mines may be of interest. Before the war steam coal was costing this company $29s.0\frac{3}{4}d.$ per ton; the price is now $50s.7\frac{3}{4}d.$ In view of the fact that it is consuming 9,000 tons per year this one item alone is now costing the company an additional £9,000 per annum. Most other supplies have increased from 12% to 50% in price and some even more. Three years ago this company crushed 6,063 tons of ore for a recovery of 76 tons of black tin, the number of men employed underground being 249 and costs were 30s.6d. per ton. Now a typical month's output has fallen to 5,114 tons, with only 56 tons of black tin recovered; the number of men working underground has declined to 158 and costs per ton have risen to 39s.11d. Were adequate developments being carried out the cost would be considerably higher and the output even lower. Such figures speak for themselves.

In addition to the factors already mentioned there are also the difficulties of obtaining machinery and supplies, restrictions on the capital market, and, finally, taxation. Questions of obtaining supplies and of raising capital are, in most cases, matters that could be solved immediately by official action. It seems, however, quite hopeless to expect the revenue authorities to revise their attitude towards companies whose existence is dependent upon the life of a wasting asset. Unless this attitude is revised, however, or unless the Government is itself prepared to supply the necessary capital for new projects it is certain that additional

Section 2: Mining and the War

production is not going to be forthcoming.

Possible Sources of Immediate Metal

An examination might now be made of sources from which increased output may be obtained. At the moment of writing it is understood that the Government is only interested in propositions that can be brought into production within 12 months or less. This limits the scope of the review to the following possible sources:

(1) Mines at present in operation.

(2) Mines fully equipped, but dormant.

(3) Relatively small or shallow mines that have been closed within recent years and dismantled because not economic at the then prevailing price of tin, but where, nevertheless, there still exist certain ore reserves or favourable prospects of being able to open up stoping ground quickly if given the necessary development.

(4) Small or shallow abandoned mines, or virgin prospects, which are thought to be worthy of rapid development.

(5) Dumps.

(6) Alluvials.

Of these six categories the first undoubtedly offers by far the best prospect of an early increase of output and possibly, the largest too, provided that adequate labour and metal prices are obtainable. It is believed that all the large producers are mining and crushing a tonnage far below their normal capacity, but if they are to respond to the National need they must be enabled to increase development greatly as well as stoping. As long as the future of a large mine such as East Pool remains in jeopardy it seems folly to suggest opening up alternative sources of supply.

Of the mines in the second class the two obvious selections are Mount Wellington and Wheal Reeth. The former property, which is in the Gwennap district, has only been worked to adit level, although trial winzes and diamond drill-holes have been put down below that point and have shown encouraging signs of an improvement of values in depth. The principal lode is a large, irregular, and very pyritic ore-body, which

Section 2: Mining and the War

has hitherto proved unprofitable by reason of the low recoverable values and the difficulties encountered in dressing. Crushing ceased about the end of 1938, but the mill and plant are standing practically intact. The ore reserves above adit are almost exhausted, but the main shaft has been sunk 40 ft. below adit level, at which point it cut the hanging-wall of the lode. If sinking were continued for a further 60 ft. and a new level established 100 ft. below adit it is very probable that within a short period a considerable tonnage of ore could be opened up and milling recommenced within a few months.

Wheal Reeth, in the Germoe parish, is likewise completely equipped, mining having ceased as recently as the latter part of 1939. This is a relatively small and shallow property and within considerably less than 12 months it should be possible to unwater the mine and do sufficient development to recommence milling, even if only on a small scale at first. The lodes here are very small, but the tonnage milled would probably yield a fair output of concentrate in proportion to the number of men employed.

Porkellis

In group three the two properties which spring immediately to the mind are the Porkellis mine, near Wendron, and the socalled "Whiteworks" mine, in Gwennap. Of all the dormant mines in Cornwall Porkellis offers, possibly, the best prospects for an early return on the capital invested. This is a mine that was literally "thrown away" in a panic during the slump in tin prices early in 1938. It is a relatively small property that has had a long, although very chequered, history, having been abandoned and re-opened many times. The last working, which commenced in March, 1934, was exceedingly successful at first, but the neglect of essential developments and a slump in tin brought the enterprise to a sudden and most untimely end in April, 1938. The plant was sold immediately and later dismantled, the only remaining piece of equipment now on the mine being the electric winder.

An independent engineer reported most favourably on Porkellis just before the abandonment and it was his opinion that the two principal

Section 2: Mining and the War

lodes would form a junction at no great depth below the present bottom level, which, incidentally, is only 90 fathoms below adit, or approximately 600 ft. from surface. The 90-fathom level was opened up from a winze sunk below the 75-fathom level and in addition to doing a considerable amount of development on both lodes at the 90-fathom a shaft rise was put up and holed to the previous sump just below the 75- fathom. The rise was stripped and timbered and the completed shaft brought into commission to the 90-fathom level only five months before the property was abandoned. Although operations were being conducted on a hand-to-mouth basis some while before the closing the late manager estimates that there are still fairly considerable ore reserves in sight and there is very little doubt that, given additional development and the immediate sinking of the shaft for at least one additional level, very considerable quantities of stoping ground could be opened up in a short while. The main perpendicular shaft, although rather small, should still be in good condition and, given the necessary labour and equipment, there seems no reason why the property should not be unwatered and brought into production once more within about 12 months. Porkellis is one of the most extreme and regrettable examples of a small and promising property being thrown away in a moment of depression because of a previous lack of development and adequate cash reserves and as such it seems worthy of the most serious consideration at the present time.

Whiteworks Mine

The small "Whiteworks" tin mine, standing a little to the north of the once celebrated Gwennap United Mines, is a further proposition that merits immediate attention at the present time. Operations on a small scale were carried on here for a year or two, but were abandoned at about the same time as Porkellis, as the price of tin was temporarily very depressed. Only one lode was worked, although there are at least two in the mine. The workings consisted of 4 or 5 levels extending at the most about 240 ft. from east to west and connected by short cross-cuts to a small vertical shaft. The bottom of the latter was only 230 ft. from surface and by reason of the free drainage provided by the

Section 2: Mining and the War

Great County Adit in this area the mine was dry and did not require any pumping plant. The bottom level, however, encountered water and had it been possible to sink deeper pumping would undoubtedly soon have been necessary.

The lode here is very unusual, consisting of a clean and easily-separated cassiterite in a gangue of soft white banded "killas" or clay-slate rock. It is exceedingly easy to mine, crush, and concentrate, and a considerable area was stoped out during the short period that the mine was in operation. The lode strikes slightly south of west and north of east and dips steeply to the north and it is stated that some of the stopes on it were over 10 ft. wide when breaking down the mineralized walls as well as the lode proper.

The actual milling recovery of tin was low, but it must be explained that all the ore mined by the late company was little more than the "leavings" of the old men who, at some very distant period, had worked the lode in a most peculiar and selective manner. The old miners had apparently picked out the richest "leaders", leaving small holes and an extraordinary ramification of small and irregular workings in every direction. The late company never reached the limits of the old men's workings either laterally or in depth, but it was generally thought that if they had had the cash resources to sink below adit level they would have been able to develop the lode below any point reached by the old men and that, in consequence, the nett tin content of the lode would have been found to be far higher than in the shallow workings above adit level, where the old workers had found it profitable to "pick out the eyes."

At the time Whiteworks was abandoned early in 1938 it was estimated that £5,000 would have been sufficient to enlarge the shaft and sink it a further 100 ft. in order to get well below adit level and in view of the prospects as they appeared then the property now seems worthy of the most serious consideration. Incidentally, although all plant has been removed, the mine is quite close to the Mount Wellington mill and the latter could be brought into immediate commission for handling the Whiteworks ore. Some modifications in the flow-sheet would undoubtedly be necessary, but the very simple nature of the Whiteworks ore would make it an eminently easy one to dress.

Section 2: *Mining and the War*

Other Prospects

Of the mines in the fourth group — i.e., small but promising abandoned mines or virgin prospects — there are so many that it is invidious to mention any particular property. Every mining man has his own especial favourite and a review of these numerous prospects would best be conducted by anyone making an investigation on the spot in conversation with the mining men familiar with their own particular districts. Amongst many others, certain small properties that come immediately to the mind are East Blue Hills, between St Agnes and Perranporth, the eastern continuation of the Vertical Lode east of the Votle shaft in old Wheal Kitty at St Agnes, some small tin lode workings south of East Charlotte in the Chapel Porth valley west of St Agnes that are thought to be a very good prospect, and, finally, the Trebartha-Lemarne mine at North Hill, on the eastern margin of the Bodmin Moor granite mass. The latter is a shallow and most promising tin-wolfram- arsenic mine situated on the killas-granite contact and seems to have been stopped largely because it was spoiling the amenities of a large estate in which it is situated and because of the damage of the arsenical waters from the mine and dressing plant to the fish in the Lynher River. In a time of National emergency all these objections can now presumably be set aside and, although the mine is admittedly far remote from the existing centres of the mining population and labour would be especially difficult to obtain, it seems worthy of an immediate investigation and re-opening.

Dumps and Alluvials

Of the remaining possible sources of an immediate supply of base metals the principal are dumps and alluvials and it will be as well to consider these jointly. At the moment of writing it is believed that the authorities favour the exploitation of these supposed sources of supply, but if that is true one is afraid that they will be disappointed. There are, possibly, certain wolfram-bearing alluvials in the eastern part of the county that will pay for further investigation and a few dumps may be found that will be worthy of treatment, but looked at from the broad point of view there can be little doubt that payable alluvials and dumps are now very definitely things of the past. A single small mine such as Porkellis is likely to produce more concentrate, and of a high grade too, than all the dumps in the County.

Other Resources

While enumerating various possible sources of an early increase of supply no mention has been made of the exceedingly promising shallow workings near the Great North Downs mine at Scorrier, as these are at present being actively prospected. Likewise no mention has been made of Wheal Hampton and Wheal Rodney at St Hilary, which are shallow mines and first-class speculations for tin, but like a hundred and one other propositions would probably require too much time to equip and unwater to be of interest in any short-term scheme. Should planning later be extended for longer periods, however, one cannot too strongly recommend these last-mentioned mines.

Finally, in long-term plans, the Polberro mine at St Agnes, so recently and regrettably abandoned, should receive serious consideration The writer is fully conversant with the developments in this recent unhappy venture and he last examined the whole of the workings the day before pumping ceased. Of the two major flat lodes in the district the top one alone was developed by the late company and it undoubtedly proved to be an uneconomic proposition. The bottom lode, however, is another very different matter, as was proved by the wide extent of the wonderful ore-shoot on the boundary that was being worked by the old Wheal Kitty company up to the great slump in 1930. One is strongly of the opinion that this great lode was actually intersected in the lower part of the Polberro shaft, but in the form of a series of small and disturbed branches that were not recognized as being the lode. Had there been the capital available to explore these branches eastward the history of Wheal Kitty might well have been repeated, in which case the mine would now have been taking her place amongst the major tin producers

Section 2: Mining and the War

of the county. The abandonment of this property is one of the most regrettable happenings in Cornwall for many years past. One feels very strongly that if the various engineers sent to examine the mine had had a better knowledge of the St Agnes district they would have presented very different reports to those which they actually submitted, on the strength of which further capital and labour were withheld.

Conclusion

In conclusion one feels it necessary to draw attention to certain dangers evident in the present official attitude towards Cornish mining. In the first place there seems to be a very real danger of endless delay and discussion by the appointment of committees and sub-committees and geological advisers to go over the ground for the hundredth time, exploring problems with which the Cornish mining man has long been familiar and which show no signs of an early solution. Quite apart from the fact that Cornwall may almost be said to have been one of the cradles of geology and that some of the world's best economic geologists have been extensively consulted on Cornish mining problems during the past 30 years, the fact remains that professional geological advice during that period has failed to yield a single discovery of value. Furthermore, it is an unfortunate fact that several very costly developments carried out within the County during this period on the strength of professional geological advice have one and all been failures. Several important discoveries have indeed been made during this time, but without exception they have all been the fruits of mining exploration and development and not the products of geological reasoning. If the authorities need the metal which Cornwall can produce they will, in the author's opinion, be well advised to appoint men of action who are familiar with the County and its problems and give them the power to act and make decisions without waste of time listening to theorists sitting in committee.

There is, in addition, a further unfortunate attitude of mind on the part of those unfamiliar with Cornwall which appears to be making itself

Section 2: Mining and the War

felt in official quarters. Many have the idea that everything in Cornish mining and milling practice is wrong and needs to be changed entirely and others entertain the fallacy that Cornish ore deposits must be worked on a large scale. Cornishmen are very familiar with both these viewpoints and have many times witnessed the failure and wreckage of "outsiders'" dreams in these respects. In Section 2 Part 2 the writer hopes to deal more fully with these oft-repeated mistakes. Meanwhile, in concluding the first part of this section, which is a hasty review of the short-term possibilities of Cornwall in a time of pressing National need, one can only express the fervent hope that those who are in official positions will approach the problems involved with a due sense of modesty and show a willingness to learn from Cornishmen already on the job.

Section 2: Suggestions for the Future

Section 2: Part II

PAST PRODUCTION AND SUGGESTIONS FOR THE FUTURE

The first part of this section was devoted entirely to the short-term aspect of the Cornish mineral industry with a view to suggesting ways and means of obtaining an early increase of output which would materially assist the country in its prosecution of the war. Pending further developments in this direction the present seems an opportune moment in which to review the industry's prospects. In order to do so effectively, however, it will be helpful to recapitulate briefly some general facts about Cornwall and the magnitude of its past production.

Cornwall is justly famed throughout the world for its mineral wealth and even in the most remote of mining fields there can be but few who have never heard of the Duchy and its multitude of mines. Strangely enough, however, although many able men have devoted an immense amount of time and talent to the investigation of Cornish mineral problems and despite an extensive literature on the subject compartively few of the present generation possess more than a superficial knowledge of the county. Among many other misconceptions there is a prevalent idea that the mines of the West of England are confined to Cornwall, although when Collins wrote his celebrated "Observations on the West of England Mining Region" he rightly included within the scope of that work the greater part of Devon and even a portion of Somerset. Nevertheless, as this authority pointed out, the mineral region is usually considered as being confined to the tract of country between the Land's End and the eastern margin of the Dartmoor granite mass and it is proposed to adopt those limits once more for the purposes of this review.

The extreme antiquity of many of the mines of this area and the re-working of various combinations of small leases or "setts" as portions of larger and later enterprises now makes it almost impossible to determine the number of individual mines, but it has been estimated that the total considerably exceeds 2,000. Of these the vast majority are situated within Cornwall itself and it is worthy of note that the number west of a

Section 2: Suggestions for the Future

line running from Newquay on the north coast to Falmouth on the south greatly exceeds the number to be found east of that line. Although the most extensively mineralized areas occur in the western part of Cornwall it is an interesting fact that some of the largest and most profitable mines were discovered farther eastward and Devonshire has the distinction of having produced the largest individual copper mine in the two western counties.

Ores Mined

A very large number of ores have been discovered and worked on a commercial scale in this remarkable field, but those of tin and copper have always predominated in importance. The ores of arsenic and tungsten, the latter especially during the present century, have been mined mostly as by-products of tin and copper lodes, as they are rarely found in payable quantities by themselves. Following these major ores, those of lead, iron, zinc, manganese, and uranium have been the most valuable commercially, and in that order of importance. The silver production, although considerable, has been almost entirely obtained from the lead lodes, of which it constituted an important by-product. Likewise much of the zinc has been raised as a by-product of lead and copper mining. The remaining metallic ores, with the exception of iron, have not been discovered in sufficient quantities to render them of more than very secondary importance.

Output

At this juncture it may be useful briefly to review the general trend of tin and copper production over an extended period in order that a perspective view of the industry may be obtained. Prior to 1882 the statistics of production are not reliable, many of the figures for the 18th and earlier part of the 19th centuries being derived from private records and

Section 2: Suggestions for the Future

estimates. Furthermore, even latter-day official records exhibit slight discrepancies and the published figures are somewhat misleading, inasmuch as the Devonshire output is included with that of Cornwall up until 1881, after which date the figures for Cornwall alone are available. Nevertheless, from the obtainable data a general estimate of production during the past two centuries can be obtained.

Mining in Cornwall commenced in depth in real earnest in the middle part of the 18th Century, consequent upon the introduction of the early pumping engines, and until the closing years of that century the output of both tin and copper had varied from approximately 2,000 to 4,000 tons of metal per annum. The relative importance of copper production then increased at a rapid rate and by 1826 the annual output had reached 9,026 tons of metal, in comparison with 4,603 tons of tin. Copper mining continued to expand so that during the period 1830 to 1860 the combined output from the mines of Cornwall and Devon was in the neighbourhood of 12,000 tons of metal per annum. Indeed, it is noteworthy that during the early part of the century the two western counties were contributing more than half the world's total output of copper ore. The peak of production was reached in 1856, when 209,305 tons of ore yielding 13,274 tons of metal were sold, the average value per ton of metal at the mines at that date being £92. By 1863, however, the decline had already commenced and the fall in output was even more rapid than the decline in price. Production diminished almost continuously and by 1886 it had dropped to only 680 tons and by the close of the century the copper industry was practically extinct.

The great expansion of tin mining did not commence until about 1823, but from then onwards it continued to increase until a maximum of 16,272 tons of black tin, yielding 10,900 tons of metallic tin, were sold in 1871. At that date the price per ton of black tin at the mines was £78 12s. 6d. During the whole period 1862 to 1893 output continued in excess of 8,000 tons of metal per annum, but with the onslaught of the great depression in the mid-nineties it declined smartly and by the end of the century had fallen below 4,000 tons. By 1913 the production had recovered to 5,288 tons, but in consequence of the last war it again declined and during the great slump in 1922 it ceased almost entirely. By 1929 the output had once more reached 3,271 tons, but the worldwide depression in 1931 reduced the figure to 598 tons. Subsequently a considerable revival again occured and up to the outbreak of the present

Section 2: Suggestions for the Future

war annual production was being maintained at about 2,000 tons of metal.

Notwithstanding the fact that the output of tin continued at a high level for many years after the copper industry had virtually become extinct the tonnage of copper produced in the West of England during the past two centuries still approximately equals the total output of metallic tin during the same period. This fact may surprise some who associate Cornwall principally with tin and who regard copper as being of quite secondary importance. Mining for tin, however, was being conducted in Cornwall long before it was realized that copper was of value and with the general exhaustion of the latter metal at relatively shallow depths tin has long since resumed its old position as the most important product of the Cornish mines.

It is hoped that the accompanying diagram (Fig.1) illustrating tin and copper production and tin prices will be of interest to those concerned with the statistical aspect of the industry. The yield of lead has not been included, as the figures available only cover the period 1845 to 1882 and as lead production has been declining since 1847 only a very partial indication can be given of the total output during the period under review. The general downward trend of tin production for many years will probably confirm the pessimists in their estimate of prospects, but before reaching hasty conclusions it is necessary carefully to study the thin continuous line illustrating the price of tin, as this demonstrates the violent price movements of recent years which have had a very disturbing effect on world tin production as a whole. Furthermore, especially as far as Cornwall is concerned, it should be emphasized that there are several other factors besides price or exhaustion of known ore-bodies that ultimately govern output.

For the information necessary for the construction of this graph the author is indebted to the Geological Survey for a similar diagram covering the period up until 1903, subsequent data having been obtained from official or other reliable publications, for the use or loan of which he desires to acknowledge his gratitude to the Editor and to H.M. Divisional Inspector of Mines at Birmingham.

Section 2: Suggestions for the Future

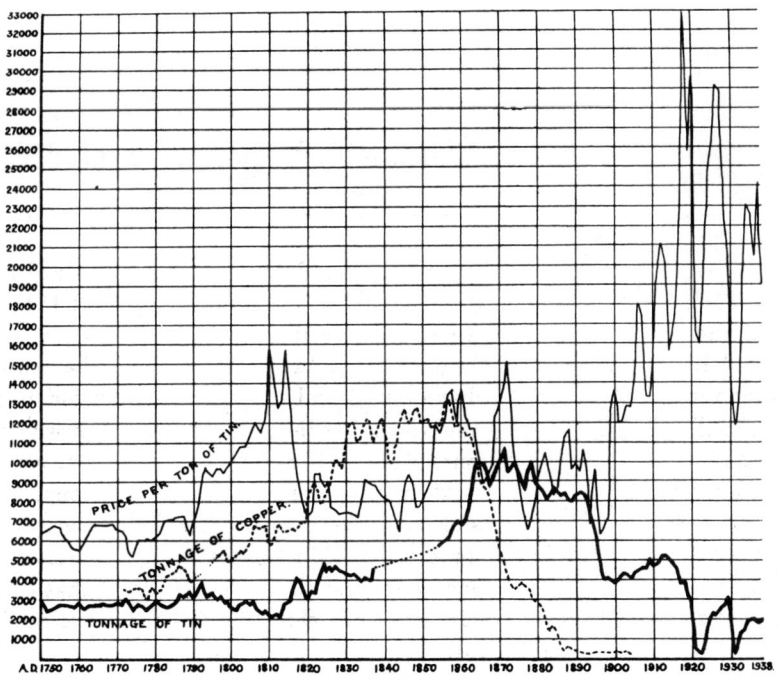

Table 1

Output of Individual Mines

Tin	Output Range	Number of Mines
Output of Black Tin	Over 20,000 tons	8
Output of Black Tin	5,000 to 20,000 tons	27
Output of Black Tin	1,000 to 5,000 tons	52
Output of Black Tin	Less than 1,000 tons	350 (approx)

Copper	Output Range	Number of Mines
Output of Copper Ore	Over 100,000 tons	24
Output of Copper Ore	20,000 to 100,000 tons	59
Output of Copper Ore	5,000 to 20,000 tons	86
Output of Copper Ore	Less than 5,000 tons	300 (approx)

Lead	Output Range	Number of Mines
Output of Lead Ore	Over 10,000 tons	9
Output of Lead Ore	5,000 to 10,000 tons	4
Output of Lead Ore	1,000 to 5,000 tons	18

A further analysis of the important tin and copper mines and those which produced both metals in appreciable quantities discloses the following important facts:-

IMPORTANT PRODUCERS OF TIN AND/OR COPPER ORES:

		No. Mines
Mines having produced at least 1,000 tons of Black Tin and/or 5,000 tons Copper Ore	Cornwall	213
	Devon	14
	Total	227

IMPORTANT PRODUCERS OF BOTH TIN AND COPPER ORES:

Mines having produced a minimum of 1,000 tons of Black Tin in addition to a minimum of 5,000 tons of Copper Ore.	29

Table 2

Major Producers of Tin and Copper Ores

Wheal Agar	South Frances
Wheal Basset	Levant
West Basset	North Roskear
Botallack and Carnyorth	Par Consols
Wheal Buller	Phoenix United
Wheal Busy	Poldice and Wheal Unity
Carn Brea	East Pool
Great Condurrow	Prince of Wales
Cook's Kitchen	Providence Mines
Creegbrawse and Penkevil	Wheal Seton
North Crofty	West Seton
South Crofty	Stray Park
Dolcoath	Tincroft
Great Wheal Fortune and others (Marazion)	Wheal Tremayne
	Tresavean (Gwinear)

Section 2: Suggestions for the Future

Scale of Past Operations

As illustrating the measure of activity that once prevailed in the West of England it is worth recording that in 1856, when copper production was at its highest level, the number of mines producing that metal in Cornwall was stated to be 134 and in Devon 23. In 1859 there were about 180 mines producing tin and in 1860 there were, altogether, more than 300 mines working for tin, copper, lead, and other ores, the number of their employees being estimated at approximately 41,000. As in many other mining fields, however, the greater part of the total output was obtained from a relatively few large producers. Furthermore, in view of the close association of tin and copper ores in Cornwall and Devon it is not surprising to discover that in many instances the largest producers of copper subsequently developed in depth into the most prolific sources of tin.

The statistics of production for individual mines are in many cases seriously defective. There are good reasons for thinking that certain mines once yielded very large quantities of ore of which no record now exists. Inasmuch, however, as recorded production is of assistance in assessing the relative importance of the tin, copper, and lead mines of the western counties, the brief analysis shown here in Table 1 may be of interest. For the interest of those familiar with the individual Cornish mines the last group consists of the 29 properties shown in Table 2. It is noteworthy that none of the Devonshire mines is included in this list. Indeed, Birch Tor mine, situated in the heart of the Dartmoor granite mass, is the only mine in Devon which has a recorded output of black tin in excess of 1,000 tons.

Geology

Having reviewed the past production of the more important ores at some length it is necessary before discussing the future to touch briefly upon the geological structure of Devon and Cornwall. Anything but the most elementary explanation will be out of place in this review, as the geology of the western counties, although basically simple, is extraordinarily

Section 2: Suggestions for the Future

complex in detail. Anyone interested in the subject is well advised to refer to the *Memoirs of the Geological Survey* and the publications of J.H. Collins, Brenton Symons, W.J. Henwood, Sir Clement le Neve Foster, Sir Henry De La Beche, and the many other writers who have dealt so exhaustively with the economic geology of the area.

The greater part of the land surface of Devon and Cornwall consists of sedimentary clay-slate rocks, which, although formed during widely varying geological periods and divisible into many groups, are known collectively to the miner as the "killas". Intruded into this mass of clay slates are five major and several lesser intrusive bosses of granite, which form the greater part of the high moorlands and hills of the West country. The principal granite intrusions from west to east are, respectively, the Land's End, Carnmenellis, St Austell, Bodmin Moor, and Dartmoor masses. Of the lesser granite outcrops the most important from an economic viewpoint are those of the Tregonning and Godolphin Hills between Penzance and Helston, the Carn Brea – Carn Entral ridge and Carn Marth – northern bastions of the great Carnmenellis mass, the small outcrops of St Agnes Beacon and Cligga Head, near Perranporth, and the Kit Hill and Hingston Down bosses, near the eastern border of Cornwall.

In addition to these major igneous intrusions, both the sedimentary and granite rocks are traversed by numerous quartz-porphyry "elvan" dykes, closely allied in composition to the granite and possessing a similar strike to the mineral-bearing lodes. The latter, incidentally, throughout the county as a whole, have an average bearing of a few degrees N. of E. and S. of W. The elvans frequently occur in the killas far from the nearest surface exposure of granite, but like the individual outcrops of the latter it is probable that they all derive their source from one common igneous mass underlying the sedimentary rocks of the entire area Indeed the elvans probably represent the dyke phase of the granite eruptions. Even older than the granite and elvans are the intensely hard "greenstone" beds and sills which intersperse the killas rocks in certain areas and which in many instances have had a very unfavourable effect upon the deposition of mineral in the lodes.

Although extensive denudation has exposed the major granite intrusions, so that they now appear as a series of dome-like elevations above the surface of the land, it is fairly certain that at the time of their formation they were covered by an immense layer of sedimentary rocks

Section 2: Suggestions for the Future

and it was under the weight of the latter that they cooled and solidified. In addition to the intense heat that radiated from the molten mass this appears to have been highly charged with intensely-active gases and solutions and in consequence the superimposed sedimentary rocks underwent extensive physical and chemical changes. The resulting altered boundary layer or "metamorphic aureole" is of widely varying thickness and composition, but is present to a greater or lesser degree wherever the killas abuts on the granite. Indeed, in some areas the presence of this aureole is the sole indication of the existence of an igneous mass which, although not exposed, cannot be at any very great depth below the surface of the land.

The earth movements preceding and accompanying the granite intrusions and the contraction resulting from the cooling both of the igneous intrusions and the heated sedimentary rocks in contact with them resulted in fractures which contain the mineral-bearing lodes of the present day. Thus far the sequence of events in the formation of the ore-bodies is perfectly clear, but the combination of factors which governed and controlled the subsequent deposition of the metallic minerals within the fissures is less certain. Were these conditions known it might be possible to predict the location and extent of the ore deposits, but failing such knowledge success in mining is still dependent on a combination of luck, good judgment tempered with experience of the particular district, and, most important of all, adequate development.

Although mining men and geologists alike are unable to predict the location or value of an ore-body with any certainty past experience has much to teach those who are willing to learn. Only too frequently during the past 40 years or more capital has been lost because those responsible for operations were apparently unable to profit from other people's experiences.

Zonal Sequence of Ore Deposition

It has long been realized in Cornwall that, as in many other fields, there is a general order of deposition of the various minerals within the lodes. The great majority of the latter having an approximately north to south

Section 2: Suggestions for the Future

strike are of a later age than the east to west veins and when mineralized have in most cases only been found to carry lead and zinc. The east to west lodes, however, have been productive of practically all the numerous minerals found within the county. Although it is rare to find all the principal minerals within a single vein they may, as a very broad generalization, be said to occur as follows:-

In the lower parts of the lodes in or near to the granite tin alone is found. In ascending order, ores of arsenic and tungsten occur with the tin and then copper makes its appearance. Still farther from the granite, first the tin and then the arsenical and tungsten contents decrease while copper increases, until finally the latter remains as the sole important metallic content of the vein. Copper is followed upwards by ores of zinc, lead, iron, and finally manganese. Denudation in many cases has removed some of the upper zones and in certain districts the process of weathering has been so extensive that only the lowest portions or "roots" of the lodes remain, in which case they yield only tin and, as might be expected, are not usually found to be productive to any great depth.

Until the middle of the last century it was generally thought that apart from the distinctive lead lodes and other veins yielding iron and other minerals of secondary importance the whole of the Cornish lodes could be classified as either tin or copper producers. It was not then realized that the copper ores would give place at greater depths to even more valuable deposits of tin within the same veins. The credit for perceiving this fact largely belongs to Capt. Charles Thomas, the then celebrated manager of Dolcoath, who strongly urged upon his company the necessity for deeper sinking. Upon adopting his pressing advice it was found that after passing through a relatively barren zone, where both copper and tin values were low, the lodes changed into phenomenally-rich repositories of tin and Dolcoath became one of the greatest, if indeed not the most productive, tin mine ever worked in the West of England. Before this change of ore contents in depth had been demonstrated by the pioneer work at Dolcoath whole areas throughout the county had been abandoned on reaching the lower limits of the copper zone, which appears to have averaged about a thousand feet in depth from surface. When the significance of the Dolcoath discoveries had been grasped the pendulum swung in the opposite direction and there was a general scramble to sink in the hope of finding tin. Since then, however, it has been fairly conclusively demonstrated that tin in workable quantities does not always

Section 2: Suggestions for the Future

succeed copper in depth and failure to realize this had been responsible for a great waste of capital.

There have been numerous illustrations of the fact that tin may occur in very large quantities in lodes situated in the killas; indeed, in St Agnes, which is a typically "tin" district, granite has never yet been encoutered in depth. Nevertheless, broadly speaking, tin is most frequently found in the lodes when in or close to the granite. In most areas where shallow copper mines have been abandoned on reaching the lower limits of the copper zone in the killas the chances of discovering tin in payable quantities below the copper are definitely good *provided that the granite is not remote in depth*. The relation of the granite relative to the bottom of the copper zone seems to be of vital importance and great attention must be given to this point when selecting areas worthy of further development. Deep and costly trials in several districts suggest that tin does not succeed copper in depth in payable quantities under the following conditions:-

(a) Where the copper zone peters out in killas far above the granite and especially where the lodes continue downward into extensive beds of greenstone, in which rock they frequently become completely barren or disordered, or even die out altogether.

(b) Where the copper zone occurs principally in the granite.

As examples of mines in the first category the Mellanear copper mine near Hayle and the celebrated Devon Great Consols may be mentioned. Both these properties belied the hopes that were entertained of them becoming tin mines in depth. A much more recent and very disappointing failure was the Tolgus mine, near Redruth, where a once productive copper lode dwindled in depth to an entirely barren and shattered vein enclosed in beds of greenstone and hard killas and as the granite proved to be exceedingly deep it was not considered worth sinking farther in order to reach it and any deepseated lodes that might have been found within it.

The failure of the Wheal Buller company at a slightly later date lends weight to the belief that where lodes are situated almost entirely in granite tin does not succeed copper at increased depth. The once-famous Tresavean mine is a further example of this contention, although in this instance a relatively shallow tin zone was encountered beneath a deep copper deposit.

Section 2: Suggestions for the Future

Depth of Ore Zones

The Dolcoath company pioneered the policy of deep sinking below the shallow copper zone and, in addition, their lodes proved to be outstanding exceptions to the general rule which seems to govern the depth at which profitable ore-bodies are to be found in Cornwall. In this particular instance the greatest riches extended down to 2,500 ft. from surface or, at the eastern end of the mine, more than 2,000 ft. below the killas-granite junction and operations ultimately reached a depth of more than 3,000 ft. from surface. The great discoveries resulting from this policy of deep mining at Dolcoath were responsible more than any other single factor for the once general belief that still lingers in some quarters that Cornish ore deposits increase in value with increasing depth. As a consequence of this erroneous idea there was a tendency, especially in the central area, in which Dolcoath is situated, to continue sinking indefinitely in the expectation of discovering better values. The too-ready acceptance of this theory resulted in a further great loss of money before it was finally realized that Dolcoath was an exception to the general rule.

One of the most important lessons to be learnt from the past is that in the great majority of the tin or combined tin and copper lodes the ores of both metals are confined to a zone whose limits do not extend for more than about 1,000 ft. to 1,500 ft. above or below the killas-granite junction. As pointed out in a previous paragraph granite has never yet been seen underground in the St Agnes district and that area may therefore be a notable exception to the general rule. It must be admitted, too, that there are numerous copper mines in many parts of the county that were worked in killas where the granite is apparently very far distant, both laterally and in depth. Nevertheless, when seeking tin, if granite is present at all within the area under examination, it is a fairly safe assumption that that mineral will not be found in payable quantities at a much greater distance than 1,000 ft. to 1,500 ft. above or below the face of the granite.

There are variations of this broad rule as, for example, the lodes of Dolcoath, which, on penetrating the margin of the granite, temporarily became hard and poor before they opened out into magnificent ore-bodies that extended to an unusually great depth. By contrast, the new East Pool mine lodes were richest immediately below the killas- granite

Section 2: Suggestions for the Future

contact and in some instances — such as, at Unity Wood and West Poldice in Gwennap — great bodies of tin and copper in the killas died out immediately on entering the granite. Nevertheless, where a lode has proved to be productive of tin in considerable quantities *in granite*, if, with increasing depth, it is found to become poor it is usually prudent to concentrate on lateral development instead of spending further capital in sinking. A more ready appreciation of this fact would have saved much useless expenditure in the past.

Lateral Development remote from Granite Outcrops

Developments in various areas having demonstrated the limits in depth of the ore zones and, more especially, the relation of the tin zone to the granite, much more attention has rightly been paid to lateral development within recent years. It has been widely argued that as the granite probably underlies the whole killas surface of the county it would only be necessary to sink sufficiently deep on the outskirts of the old-established mining areas in order to intersect additional lodes in or close to the margin of the granite. It was hoped that vigorous development of such lodes in this zone would lead to the discovery of further important reserves of tin. In support of this idea it was suggested that the numerous lodes that outcrop at the periphery of the old mining areas and which contain small quantities of the upper-zone minerals — such as zinc and copper — would prove in depth to be deep-seated copper and tin lodes similar to those so extensively exploited close to the granite bosses. This line of reasoning gave rise to the various schemes for developing the "Northern Areas" of the central mining district between Redruth and Camborne. Unfortunately, however, deep and costly developments at Tolgus, East Pool, South Crofty, and Roskear have done much to discount the truth of this theory. It appears that with increasing distance from the granite outcrops the ore-bodies become progressively smaller, more widely separated, more irregular, and of lower average value. Admittedly such experience in one district is far from being conclusive evidence that negative results would also be obtained by similar developments elsewhere.

Section 2: Suggestions for the Future

Up to the present, however, the available evidence is strongly suggestive that in Cornwall the major ore-bodies, and especially those yielding tin, are to be found relatively close to the granite outcrops, or where that rock is at no great depth below the surface of the land.

Past Failures

Cornwall has had more than its fair share of thoroughly discreditable "promotion" flotations during the past generation and all too frequently capital that ought to have been devoted to the development of a property has been paid away in the form of iniquitous "purchase considerations". In a relatively few instances, of which the recent Polberro enterprise is a distressing example, a genuine venture has failed through shortage of capital necessary to carry it through to success. There have been other cases, too, where the wrong type of plant has inadvisably been installed and where a succession of mechanical troubles culminating possibly in a serious breakdown (as at Wheal Vor) have exhausted the capital resources of the operating company before any new developments worthy of note could be carried out.

It must also be admitted that there have been a few, although only a very few, well-conceived and wisely-conducted schemes that failed because the mines themselves were poor. In spite of these latter disappointments it may truly be said that the majority of the failures of the past 40 years have been brought about by re-opening mines which a little common sense alone would have indicated as being unfavourable prospects. The basic reason underlying the flotation of most of these ill-advised ventures has been the idea that because a mine was once very productive for tin it is likely to be so again in the future, whereas in all probability it is almost completely worked out.

In a very general way the tin and copper mines of Cornwall may be divided into the following three groups:-

(1) Copper mines which have exploited the shallow upper zones of the lodes situated mostly in killas.

(2) Tin mines worked in the deep zones, mostly in granite, which through extensive denudation is now exposed at surface.

Section 2: Suggestions for the Future

(3) Copper and tin mines where, following the exhaustion of the copper at relatively shallow depths in killas, tin was discovered by additional sinking usually into or close to the granite below.

As far as the second and third groups are concerned, unless there is some particular reason for thinking to the contrary, all those seeking new discoveries would be well advised to leave them severely alone. It is through re-opening such properties that the majority of past failures have occurred, even if there are a few instances where other causes have led to the closing of a mine which still thoroughly merits re-opening. In the majority of cases, however, the fact must be plainly faced that where the previous generations worked a mine on a considerable scale for tin, either alone or subsequently to the shallow copper deposits, there are but slender hopes of making further payable discoveries. It is surely obvious that the old men, like the present generation, continued to work and explore a piece of ground as long as they thought there was the slightest prospect of it remaining or becoming profitable. Furthermore, they were just as astute and far more capable than present-day mining men in the art of working successfully the relatively small and complex mineral deposits that constitute the majority of the Cornish mines. It must not be forgotten either that although the price of tin was on the whole very much lower during the 18th and 19th centuries than that prevailing during the present wages and all costs were, nevertheless, proportionately low and former generations were frequently able to prosper just as much with tin at £100 per ton as present-day companies can do with the price at double or treble that figure.

It is an oft-repeated but fallacious argument that the "old men" could not handle their ore-dressing problems as well or as cheaply as can be done to-day, whereas with the exception of the invention of the magnetic separator, which solved an otherwise very difficult problem for those mines which possess wolfram in their ore, we are probably no better and no worse off than were the old men as far as concentration is concerned. What we now do by machinery they did by hand and with very cheap labour at that. Therefore, from every point of view, unless a detailed knowledge of the history of a mine provides good and convincing reasons why it should be re-opened it is better left entirely alone, if in the past it has been extensively worked for tin.

Some may object that an entirely new mining or ore-dressing process — such as "sink and float" — may be applied to the working of the

Section 2: Suggestions for the Future

Cornish ores and revolutionize working costs, thus making available low-grade ores that cannot possibly be profitably mined at the present. Such a development may well occur one day, but meanwhile it is just as well to bear in mind the failure of several companies that were sponsored in the belief that present-day methods could succeed where old ones had failed. The unfortunate shareholders in these concerns discovered to their cost that what the old men had left as being unpayable is equally useless to the present generation. It has truly been said that "One cannot sell what was sold 60 years ago!"

Prospects

So far attention has been devoted entirely to the pitfalls that may be encountered when seeking new mineral deposits. The numerous individuals who have burnt their fingers badly in Cornwall in the past will probably consider that what has already been written merely emphasizes their contention that the mineral areas of the West of England are practically exhausted and as such no longer offer worth-while opportunities for further extensive development. The writer, however, is unaware of any facts that justify such a pessimistic opinion, but if further serious mistakes are to be avoided it is essential that the reasons underlying past failures be investigated and emphasized.

In the next section it is intended to review some of the individual districts and there mention will be made of certain mines and virgin prospects that merit development. At the present juncture only general principles are being considered and it will, therefore, be convenient if the writer summarizes his own views of the future as follows:-

(1) With but few exceptions the ores of the shallow lode zones — such as, copper, lead, and zinc — have been most extensively developed and are practically exhausted and unless some entirely new discoveries are made they do not merit further investigation.

(2) Iron does not occur in sufficiently massive deposits to make its exploitation a commercial proposition except in times of exceptional scarcity and unusually high prices.

(3) Arsenic, being basically a by-product of tin and copper mining, and wolfram, like-wise being mined mostly as a by-product of tin, the

Section 2: Suggestions for the Future

future production of both these ores will be largely governed by the activity of tin mining. It is probable that fairly extensive reserves of each of these minerals exists and especially is this true of wolfram, which is found to occur with tin in at least two very extensive though low-grade stockworks.

(4) The contradictory statements often heard that "Cornwall is not yet scratched" and that "Cornwall is worked out" are equally far from the truth. As in other of the world's mining fields many of the best and most easily-worked ore-bodies have admittedly been exhausted and although it is quite unreasonable to expect a return to an era of activity comparable to the palmy days of the last century there are probably considerable reserves of tin yet awaiting discovery and profitable exploration.

(5) Although the possibility of exceptions to the general rule must always be borne in mind, the most favourable areas in which to seek for tin are the shallow copper-mining districts *where there are good reasons for thinking that granite exists at no great distance below the bottom of the copper zone*. Any evidence of the presence of arsenic and more especially wolfram in such mines is a further strong indication of a probable change from copper to tin at greater depths.

(6) With but few exceptions the best deposits of tin are found close to or within the margin of the granite. When following a lode downward into the latter rock it must always be borne in mind that the depth of the tin zone is limited and if values fall off it is wise to seek for parallel ore-bodies rather than to follow the existing ones still deeper.

It will have been noted that as yet no mention has been made of the possibility of discovering entirely new mineralized areas. That such may one day be found is not beyond the bounds of possibility, but the prospects of making such discoveries do not seem to be very encouraging. Wherever mineralization exists on an extensive scale one or more lodes are almost certain to outcrop at surface and in the past, when the land was far less extensively cultivated than is the case to-day, such ore-bodies were far more readily detected. Therefore, in view of the fact that the whole county has been most diligently prospected for centuries, there seems little likelihood of discovering altogether new mining areas. On the whole it seems much more probable that most future discoveries will be confined to the established mining districts and that they will be obtained by deeper sinking in the shallow mines and by more extensive

Section 2: Suggestions for the Future

lateral development in order to intersect parallel ore-bodies. Incidentally, the lack of adequate cross-cutting in many mines is the most serious criticism that can be made of the old men's methods. This failure to explore for new and additional lodes while exploiting known bodies has been one of the most serious and frequently repeated errors in the whole history of Cornish mining. Indeed, the prospect of intersecting other and entirely unworked parallel lodes in depth offers a certain amount of justification for re-opening some mines that otherwise would not merit a moment's further consideration. In this connexion the story is told of the mine manager who, when asked why more cross-cutting had not been carried out in the mine of which he had once been the underground agent, replied "Well we had a good mine and we didn't trouble to look for another."

Conclusion

"Can any good come out of Cornwall?" asked a sceptic a few years ago. "No," was the cynical reply, "if they have anything good they keep it to themselves." Such remarks are eloquently expressive of the popular opinion of Cornish mining within recent years and in view of much that has occurred the widespread public prejudice that exists is not in the least surprising. The writer, however, is not alone in thinking that if in the future Cornwall is tackled with a more intelligent appreciation of the mode of occurrence of its ore deposits and with a greater willingness to profit from past experience then a very considerable number of highly-successful new tin-producers will be brought into existence. In the next section it is proposed to indicate some of the more promising areas for such future exploration.

Section 3: St Just and St Ives Districts

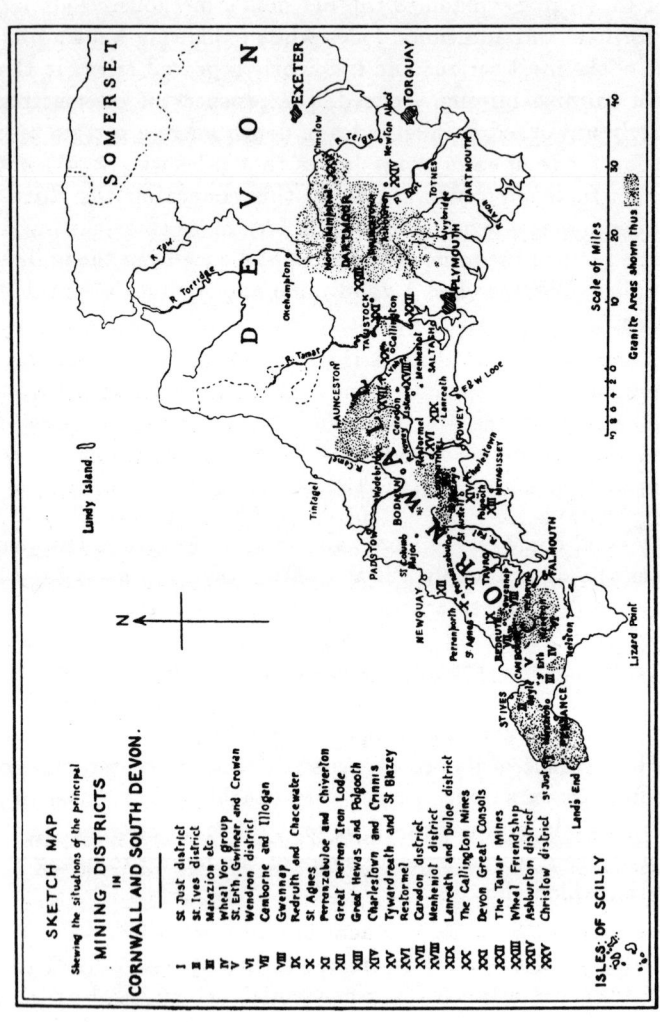

SECTION 3: REGIONAL SURVEY OF THE MINING DISTRICTS

Section 3: Part I

THE ST JUST AND ST IVES MINING DISTRICTS

In the second part of Section 2, the production of tin and copper from West country mines during the past two centuries was reviewed and a number of suggestions made concerning the principles that should be observed when seeking new deposits, more especially those of tin. It is now proposed to describe briefly the principal mining districts with their more important features and some of the outstanding mines in each area. Particular mention will be made of certain of the old properties and the more or less virgin areas that offer the best prospects for the discovery of new and important ore-bodies.

THE ST JUST DISTRICT

Reference to the sketch map opposite will show that the first district of importance in the western part of Cornwall is situated on the coast within a few miles of Land's End. Although the district derives its name from the town of St Just, the active mines in recent years have been almost wholly confined to the northern end of the area. The St Just district consists of a mineralized belt which extends along the coast for a maximum distance of about four and a half miles, although it is rarely more than a mile wide and frequently less. Notwithstanding its relatively small area it has been one of the most important mining districts in Cornwall and over 50 lodes are known to exist within three and a half

Section 3: St Just and St Ives Districts

miles of country. Basically the district consists of a coastal strip of mixed greenstone and highly-altered killas, about a third of a mile in width, which borders the north-western margin of the Land's End granite mass. The lodes occur in the granite as well as in the killas and greenstone, but the majority of the richest and largest mines have been worked in the sedimentary rocks or in the margin of the granite and in consequence the mines with but few exceptions are situated within less than a mile of the coast.

The lodes of this district have frequently been found to contain ores of both copper and tin, even at the greatest depths reached, and although narrow — Henwood estimated their average width at only 2.23 ft. — are on the whole above the average grade of the Cornish ores. The lodes are mostly steep, 70° to 80°, but their outstanding feature is their general north-westerly strike — indeed, they run almost at right angles to the majority of the Cornish tin and copper lodes. Furthermore, as they cross the killas- granite junction at right angles and as the igneous rock underlies the slates seawards and the ore-shoots tend to follow the contact, many of the mines have been worked far out under the sea. In some of the most important properties — such as Levant — it has been found that the ore values are almost entirely confined to those parts of the lodes that are situated in the killas and greenstone; at quite shallow depths in the granite the lodes become small and poor. Consequently, in order to follow the rich ore-shoots at and above the contact, it was necessary to adopt elaborate and costly submarine methods, which greatly increased working costs. In compensation for this disadvantage, however, the mines are singularly free of water and even at Levant, where the workings extend for a mile and a half under the Atlantic, the total quantity of water pumped was only about 60 gallons per minute.

Although the St Just district contains upwards of 40 mines there are only five whose recorded output has reached major proportions and a few details of these mines may therefore be of interest.

Levant

The first mine of note and, indeed, one of the most famous in the whole of Cornwall is Levant. After an earlier and unsuccessful working operations

Section 3: St Just and St Ives Districts

there were recommenced in 1820 and from that date were continued without a break until the autumn of 1930. During this period the sum realized by the sales of tin, copper, and arsenic exceeded £2,800,000. It is recorded that within the first 44 years the fortunate shareholders received for an outlay of £7 10s. per share £1,091 in dividends and the mine continued to be very profitable throughout the greater part of her long career. Had it not been for the great disaster in 1919 which resulted in the death of 30 men and the very unfortunate mining and financial policy that was subsequently pursued, it is highly probable that Levant would still have been on the active list. In spite of the writer's previous condemnation of the practice of re-opening old and extensively-worked tin mines it must be admitted that Levant is in a category by itself and it would seem that there are few mines in the West of England that are more worthy of the cost of re-opening.

There are numerous lodes in this wonderful old mine, some of which, incidentally, have been quite inadequately explored in the deeper levels, but the major riches have been discovered in the vicinity of the junction of the North and South lodes. The former bears W. 28° N. and dips steeply south and intersects the north-dipping South lode, which has a bearing of N. 40° W. The great ore-shoot north-west and south-east of the point of intersection is over 3,000 ft. in length and has been extensively worked to a depth of 350 fathoms below adit level. The grade of ore was far above that of most of the other Cornish mines working during the past quarter of a century and indeed a director long associated with the mine has told the writer that at the lowest possible estimate the deeper workings will average 37lb. of black tin per ton.

As previously explained, the ore values are mostly confined to those parts of the lodes situated in the killas and, in consequence of the deeper portions of the main shafts near the cliffs having entered the barren zone in the granite, extensive and unproductive drives on the strike of the lodes had to be extended seawards at each level before payable values were again encountered. To avoid similar development at greater depths in the days of hand drilling a subsidiary shaft was sunk below the 210-fathom level at a point over 1,000 ft. seawards of the land shafts. In spite of the very considerable natural heat of the mine, hoisting in this — the Old Submarine shaft — was performed by means of a steam engine and boiler erected in old workings. Even more remarkable was the introduction underground of a small steam locomotive with a view to

Section 3: St Just and St Ives Districts

reducing haulage costs, but the heat and fumes quickly compelled the abandonment of this experiment. The Old Submarine shaft was sunk to the 302-fathom level, but at a later date the excessive cost of rehandling the ore necessitated the further sinking of the land shafts, which were accordingly deepened to the 278-fathom level at which horizon very long drives were extended seawards. Finally, a further subsidiary shaft was sunk approximately 2,500 ft. from the coastline and from this shaft the workings were extended to a depth of 350 fathoms. However, working costs remained exceedingly high, as the ore from the bottom western stopes had to be rehandled about seven times before it reached the mill and all pumping and hoisting from the deep levels was performed uneconomically by means of compressed air - e.g., the large direct-coupled winding engine at the New Submarine shaft alone absorbed the greater part of the output of a very large air-compressor. The workings under the sea ultimately extended for a distance of one and a half miles from the land shafts and so much of the men's time was occupied in getting to and from their working places that only five out of eight hours were available for actual work. Furthermore, as the air currents had to traverse such a circuitous route the mine was very hot — over 90° — and the greater part of the drilling continued to be done by hand. In view of all this it is not surprising that at one period during the last war working costs reached a peak of 54s. a ton.

Towards the end of 1919 negotiations were in progress for the conversion of the old "cost book" or unlimited liability company to a limited one with a view to raising further capital and sinking a new main shaft to give direct access to the bottom of the mine. In October of that year, however, a disastrous accident occurred to the "man engine", by means of which the miners were still raised and lowered, and this resulted in the death of 30 men and injury to many more. At this time the company were carrying their own accident insurance risk and the crippling burden this imposed together with the loss of all means of direct access to the bottom of the mine very nearly brought the enterprise to a close. A financial reconstruction of the company was eventually achieved early in 1920, but in consequence of a breakdown of negotiations with the interests who had originally intended underwriting the capital issue only a fraction of the money required was forthcoming and it proved impossible to sink the new shaft that was so urgently required.

Following these misfortunes an unhappy decision was reached as

Section 3: St Just and St Ives Districts

to future mining policy. Instead of equipping the Man Engine shaft with a man hoist, as a temporary measure, which would have permitted the continued working of the bottom levels it was decided to abandon the lower half of the mine and to concentrate all development in the upper levels. During the next ten years all work was confined therefore to the part of the mine between the 130 and 190-fathom levels and on surface a new mill and much other new plant was installed. The old mine responded to this development in a wonderful manner and from these old shallow levels, hitherto thought to be exhausted, over £385,000 worth of tin was recovered. The limited company, however, was in financial difficulties from its inception and with the onslaught of the world-wide depression in 1930 the final collapse came and thus ended, at least temporarily, the working of one of the most productive and profitable mines ever discovered in the West of England.

The actual tonnage of mineral produced during the limited company's existence is not available, but the following figures covering the 100 years of "cost book" company operation may be of interest:-

Copper ore, 136,562 tons realized	£859,303
Tin ore, 21,959 tons realized	£1,531,977
Arsenic soot, 3,800 tons realized	£32,657
	£2,423,937

Whatever mistakes may have been made by the limited company it was the failure of their predecessors to sink a new shaft during periods of prosperity that was ultimately responsible for the untimely closing of Levant. Had such a shaft been provided the man engine would have been displaced long before the disaster occurred, the air temperature would have been reduced, an enormous saving of time in reaching the various working places would have been achieved, and in every direction a drastic reduction of working costs would have been brought about. Incidentally there has been a considerable difference of opinion as to the relative merits of an inclined as opposed to a vertical shaft at Levant. In spite of the many advantages of the vertical it seems that in these particular circumstances the inclined shaft is greatly to be preferred and if ever Levant is reopened it is to be hoped that such a shaft will be sunk.

Almost alone among the abandoned mines of Cornwall Levant may reasonably be regarded as more of an investment than a mining specu-

Section 3: St Just and St Ives Districts

lation. The sudden stoppage and abandonment of the deep and highly productive workings was brought about by a calamitous accident and not by reason of any falling off of values. The length of the main oreshoot is considerable and the grade of ore unusually high. There seems to be no indication of any decrease of values in depth and it is noteworthy that copper and arsenic in considerable quantities were present in the deepest and most westerly workings, thus indicating that the mine had not even reached the deepest zone in the lodes where tin alone is encountered. In this connexion it is not without interest that at the deepest point in the mine — a sump winze about 15 or 20 fathoms west of the New Submarine shaft — the lode carries appreciable copper values over a width of about 3 ft. Finally it should be remembered that the volume of "coming" water to be handled is negligible. Although the provision of a new shaft capable of working the property economically to greater depths would be an expensive matter there is every reason for thinking that it would prove a profitable investment. As a purely personal expression of opinion the writer most earnestly hopes that the neighbouring Geevor company will one day undertake this task before their own mine commences to show signs of impoverishment in depth.

Geevor

The second mine of importance in the St Just district, from the point of view of tin output, is undoubtedly Geevor. Since the closing of Levant this has been the only mine in operation in the area. The present leases include several mines or "setts", but the property originally consisted of the old North Levant and Geevor mines. These are very old properties that were re-opened in 1853 and worked on a small scale until about 1890, when the deeper workings in North Levant were abandoned, partly in consequence of insufficient pumping power. Operations on a small scale were, however, carried on almost continuously and in the early years of the present century a small local syndicate was formed with a view to developing the property more vigorously. Great success attended these efforts to resuscitate a mine which had always been considered very promising and, in consequence, the present "Geevor Tin Mines, Ltd.",

Section 3: St Just and St Ives Districts

was registered in February, 1911. From 1860 to 1910 the recorded output of these mines was 4,026 tons of black tin, since when an additional 15,381 tons had been sold up to the end of March, 1941, making a grand total of 19,407 tons of concentrate.

As in the case of Levant the lodes have a general north-westerly bearing, are small, of considerably higher average value than the majority of the Cornish ores, and the mine is likewise singularly free of water. Whereas, however, the riches of Levant were found in the killas rock and mostly under the bed of the ocean, the lodes of Geevor are almost entirely confined to the granite and are mined at a considerable distance inland from the coast. The only part of the mine situated in the killas is the extreme north-west extension of the main or North Pig lode and on entering the sedimentary rocks the lode soon declines in value. There are at least seven lodes of importance in the mine, some of which cross others at a considerable angle and in certain instances produce faults of some magnitude.

It appears that the ore values in Geevor, as in so many other Cornish mines in the past, are paralleling the killas-granite contact and as the mine has now reached an appreciable depth in granite it is a matter of considerable general interest how much further the lodes will continue to be productive in depth. Whatever happens in that direction the company still has very extensive opportunities for further lateral development, which, together with the possibilities of ultimate extension westward into Levant, will, it is to be hoped, long enable Geevor to maintain her position as the leading Cornish tin producer and one that to date has had a very satisfactory dividend record.

Botallack

The various mines comprising the Botallack group constitute one of the largest and at one period the most celebrated of all the St Just producers. In addition to the old Botallack mine itself the group includes amongst others the Crowns section, which was worked beneath the sea for 500 or 600 yards, Wheal Cock, which was also worked beneath the sea, and the Carnyorth mine. It is not possible to give accurate figures for the

Section 3: St Just and St Ives Districts

production of the group as the returns from the individual mines are somewhat confusing and in view of the great antiquity of some of the workings it is obvious that the statistics of production are incomplete. Nevertheless, published figures show an output of black tin exceeding 15,000 tons and of copper ore 22,465 tons.

There are numerous lodes in the combined setts which, like Levant, are situated on the killas-granite junction, but in this case they were found to be highly productive for tin in the granite, while the bulk of the copper was obtained from the seaward extension of the ore-bodies in the killas and greenstone. The lodes vary in strike, but in general they have the typical north-westerly trend of the St Just area and, with the exception of the Crowns or Narrow lode, which varies in width from 2 to 12 ft., they are all small. A great variety of minerals have been found in these veins and although narrow they were remarkably rich in both copper and tin.

As at Levant the general trend of the ore shoots followed the killas-granite junction seawards and in order to avoid further unproductive sinking of the existing shafts and subsequent lengthy and costly drivage at each level to reach the subterranean ore values it was decided to sink a diagonal shaft at an angle of $32\frac{1}{2}°$ from the horizontal across both dip and strike of the Crowns lode. This notable shaft, commenced in the greenstone in 1858, occupied four years in sinking and by reason of the excessive hardness of the rock was only made 6 ft. square, its small dimensions prohibiting the use of more than a single 16-cwt. skip. By the time it reached the bottom of the mine, then 180 fathoms deep, it had been sunk an inclined distance of 2,070 ft. and it intersected the 180-fathom level at a point 1,560 ft. seaward of the previously-existing hoisting shaft.

The remarkable situation of the engine houses and machinery, which seemed literally to be perched on ledges in the face of the precipitous cliffs, together with the great extent of the subterranean workings, combined to make this one of the "show" mines of Cornwall during the last century. In the shallow levels the old men worked in some places to within little more than 3 ft. of the floor of the sea and in one case a drill hole actually tapped the sea bottom, the nearness of which was not realized owing to the dryness of the working place, although it was stated that stones could be clearly heard rattling on the sea bottom in stormy weather. Amongst the illustrious visitors to the workings were

Section 3: St Just and St Ives Districts

Queen Victoria, in 1846, and King Edward VII, when Prince of Wales, in the '60s. Not least of the remarkable associations with Botallack was the blind miner who worked there at one period and who was sufficiently proficient that by his labours he was able to support a family of nine children. It was this man's proud boast that he had never asked for public assistance from the parish — a typical example of that sturdiness and independence of character that made the Cornish miner famed throughout the world.

Notwithstanding the fact that Botallack had been worked in places very close to the sea bed and for a depth of about 240 fathoms below sea level it was, like its near neighbour Levant, a remarkably dry mine, owing to the impervious nature of the hard killas-greenstone rocks in which its subterranean ore-bodies occur. On the whole Botallack was a very profitable mine throughout the greater part of its long history, but in latter years it seems to have lacked energy in direction and, becoming impoverished, succumbed to the great depression of the '90s and closed down in 1895.

In 1907 the mines were re-opened on an extensive scale and much money was expended on their equipment, unwatering, and further development. From the commencement, however, the wisdom of the programme of development adopted was highly questionable and when the mines were again abandoned in 1914 it could truly be said that whatever prospects of success had existed at the time of the re-opening remained entirely unexplored and untried. As has so frequently happened elsewhere, instead of trying to open up new ground under the old copper deposits those responsible made the fatal mistake of expecting to find tin where the old workers had mined it extensively in the past. If an inclined shaft had been sunk capable of efficiently developing the deep seaward extensions of the Crowns mine and Wheal Cock, which were the predominate copper-producing parts of the mines, it seems very probable that a new and successful tin mine would have been opened up under the sea. Instead of attempting such developments the new company concentrated all their energies on unwatering and exploring the landward end of the mines which had already been worked far into the granite by the old men and, as might have been expected, they drew a complete blank! A large new vertical shaft was sunk to a depth of about 1,500 ft. and the new company ultimately holed up under the bottom of the old workings, but they found that the lodes had become mere knife edges in

Section 3: St Just and St Ives Districts

depth and that the old men had completely exhausted the landward end of the mine. One or two small blocks of ore were discovered and worked out and a considerable tonnage of dumps was milled, but on the whole the enterprise was a complete disappointment. Far too much capital was expended in surface erections, including the installation of a large mill, and there were repeated changes of management and policy. By the time it had become evident that the seaward extension alone should have been tackled the capital had been spent and those responsible did not see their way to go any further with the venture. The property was therefore once more abandoned.

As a purely personal opinion the writer doubts whether the prospects in the Crowns and Wheal Cock justify the cost of once again re-opening these mines and it is obvious that the chances of success are not to be compared with those offered by the resuscitation of Levant. Nevertheless, had these mines been more intelligently handled in the period from 1907 to 1914 it is quite possible that they would again have become important producers. It is regrettable that so much capital should have been literally thrown away in demonstrating once again the futility of the idea that where the old men have mined for tin on an extensive scale workable values still remain to be exploited.

Wheal Owles

Immediately south of the Botallack group occurs Wheal Owles and the lesser mines, such as Boscean, that have at various times been worked in conjunction with it. In this case also there is a certain confusion of the records of production. In addition to a small quantity of copper Wheal Owles is recorded as having sold 8,540 tons of black tin and Boscean a further 2,400 tons and therefore the group ranks amongst the major producers of this area.

These mines have also been worked in the granite and at and about the killas-granite junction seawards. A large number of ores — including those of bismuth, cobalt, and uranium — have been discovered and marketed. These mines were very economically and skilfully managed and although the lodes were not rich they gave a splendid return over a long

Section 3: St Just and St Ives Districts

period of years. On January 10, 1893, the mine was suddenly drowned by accidentally holing to the old workings of Wheal Drea, the disaster resulting from neglect, when surveying, to compensate for annual magnetic variation; 20 men lost their lives in this accident and the bodies were never recovered. Proposals were put forward for reconstructing the company and unwatering the mines, but the great depression of the '90s discouraged any further action and although the Wheal Owles sett was included in the leases of the newly-formed Botallack company in 1907 nothing further was done and the mine remains to this day exactly as she was on that fatal day in 1893.

Balleswidden

About a mile to the east of the town of St Just and upwards of two miles from the coast is the isolated sett of Balleswidden, which is estimated to have produced 12,000 tons of black tin between the years 1837 and 1873. This property was worked to a depth of 178 fathoms from surface and is situated entirely in the granite. There are several lodes with the prevailing north-westerly strike of the area, the main lode — known as the Awboys — consisting of a series of small parallel veins which, together with intervening mineralization, constituted a deposit that was worked away for a width varying from 10 to 30 ft., the average being about 12 ft. Each little vein had a hard wall 2 in. to 6 in. in width, but beyond the wall the granite was kaolinized. In consequence the ground required much timber and accidents occurred very frequently. In recent years the site of the mine has been largely obliterated by extensive workings for china clay.

Although the other and numerous mines of the St Just area have produced in the aggregate a very large tonnage of mineral there are no others that merit especial mention in this review. With the exception of Levant and the extensive areas already leased to the Geevor company it is difficult to see where prospects exist in this area that warrant the expenditure of further large sums of money. An old mining man with much experience of the district has suggested to the writer that a better prospect than any further exploration in the immediate vicinity of St

Section 3: St Just and St Ives Districts

Just or Pendeen would be the driving of a deep adit from the base of the cliffs at Bosigran Castle, about 3 miles north-east of Levant. Such an adit, if extended for three-quarters of a mile, would, it is suggested, provide free drainage to depths ranging from 300 to 800 ft. and would enable cross-cutting and exploration in depth to be carried out cheaply on the group of lodes of the several small but hitherto but little developed mines in this neighbourhood.

Ding Dong

On leaving the St Just district and before describing the mines of St Ives and Lelant brief mention should be made of the isolated but extensive old mine of Ding Dong, which is situated in the heart of the Land's End granite mass about 2 miles north of the road running between St Just and Penzance.

Ding Dong is probably one of the oldest mines in the whole of the West of England and the details of its early history are lost in the mists of time. The last period of working extended from 1814 to 1879 and as the only records of production, covering the years 1855 to 1878, show an output of 2,905 tons of black tin it is apparent that the total output must have been considerable. There are said to be 22 lodes in the mine, the more important of which bear E. 25° N. to E. 30° N., these being intersected at approximately right angles by other lodes which sometimes fault them. Hunt says of this mine:-

The lodes were constantly throwing out branches and disseminating strings to such an extent as to appear to fill the granite with mineralized veins. No lodes come to the surface...strings run irregularly through the granite in all directions; sometimes a vertical lode is met with, but it lasts but a few fathoms and usually entirely dies out.

When Ding Dong was finally abandoned in 1879 it had reached a depth of 138 fathoms from surface. In 1926 proposals were made for re-opening, but as support from the public was not forthcoming nothing further was done. The writer cannot help thinking it was exceedingly fortunate that the attempted flotation failed, as in all probability it would have been found after unwatering the mine that it was completely exhausted and the old men had so extensively explored the ground that there was little opportunity for making any further discoveries. As in so many other cases it was obviously hoped to refloat Ding Dong on the strength of past production, none of which, however, is of any monetary value to the present generation!

THE ST IVES DISTRICT

With the exception of a few small, scattered, and hitherto unimportant mines along the northern coast no further mineralized areas of importance are to be found in the Land's End peninsula until the neighbourhood of St Ives is reached. The mineral wealth of St Ives and of the neighbouring parishes of Lelant and Towednack gave rise to what was once an important mining district, the principal product of which was tin.

As in the case of St Just the St Ives district contains a coastal fringe of killas and greenstone bordering the granite mass, but at St Ives the majority of the important mines were worked within the granite and, as such, were almost wholly stanniferous. The few mines near the coast that were worked in the killas-greenstone rocks yielded mainly copper, with which considerable quantities of pitchblende were sometimes associated. The lodes of this area were on the whole very small; Henwood estimated their average width as being only 1.61 ft. or 0.62 ft. less than those of St Just. Unlike the veins of the latter district, with their general northwesterly strike, the St Ives lodes have an average bearing of several degrees south of west — similar to those of the Camborne and Illogan area. Although small the lodes were on the whole fairly productive and the amount of water to be pumped was usually very little. The feature which particularly distinguishes this area is the "carbonas" or enormous irregular pockets or masses of tin ore which occur in the granite country rock. Apparently associated with the lodes the carbonas are frequently only connected to them by small pipes and strings of ore. The location of these irregular deposits has hitherto been so much a matter of chance

Section 3: St Just and St Ives Districts

and accident that it is possible that many more yet exist, whose presence was quite unsuspected at the time the mines were in operation.

St Ives Consols

Possibly the most famous of all the mines in this district is the celebrated St Ives Consols, which occupies the central position in the northern group of properties. The mine lies within the granite close to the killas — indeed the latter overlies the granite at the eastern end of the property. It is recorded that the old men had worked these lodes prior to 1818, but it was in that year that the mine was restarted for, peculiarly enough, purely electioneering purposes! So great was the success of a venture that had been started without any intention of *bona fide* mining that it continued in operation until 1873, by which date it had reached a depth of 197 fathoms from surface and had sold over 15,000 tons of black tin.

The principal lode is a rather narrow and almost vertical vein known as the Standard lode, but the mine owed much of its productiveness to the extraordinary series of carbonas that branched off from the southern side of the Standard lode below the 57-fathom level. Some of the excavations on the Great Carbona exceed 70 ft. in height and width and very heavy timbers had to be employed to support the ground. This section of the mine was ultimately ruined by a disastrous fire that continued to burn for six weeks and completely destroyed the whole of the timber supporting the workings. Collins states that the ore obtained from the carbonas averaged about 34 lb. of black tin per ton of stuff.

As in very many other Cornish mines the lodes of St Ives Consols became hard and unproductive on reaching a depth of 1,000 ft. in the granite and in consequence the deep workings were finally abandoned about 1873. In 1908 the mine was re-opened and partially unwatered by a group attempting to re-work several mines in the neighbourhood, but, as might have been foreseen, the property proved to be completely exhausted down to the depth to which the water was lowered and had the unwatering been completed there is little reason to suppose that more favourable results would have been obtained. The re-opening seems to have been undertaken for two principal reasons: first, it was hoped to

Section 3: St Just and St Ives Districts

discover additional carbonas, which in view of the difficulty of locating them by any known means was certain to be a very hazardous venture, and, secondly, in the hope of finding good values further eastward in the neighbourhood of the killas-granite contact. The latter prospect seems to have offered the sole justification for any re-opening of the Consols and it is to be regretted that difficulties resulting from the outbreak of the last war compelled the abandonment of the property before any extensive and conclusive developments in this direction could be undertaken.

Wheal Trenwith

Eastward the lodes of St Ives Consols pass into the killas-greenstone rocks, where they have been worked in the celebrated old Trenwith copper mine under the town of St Ives. Much pitchblende was present with the ore and it constituted a serious impurity at the time the mine was being worked for copper, but in 1909 the property was re-opened by the St Ives Consolidated Mines, Ltd., expressly for its pitchblende content. The workings were unwatered to within a short distance of the bottom level, which is about 135 fathoms from surface, and a good deal of pitchblende was obtained, especially from the dumps on surface. Unfortunately, however, the war-time difficulties that caused the abandonment of St Ives Consols also compelled the closing of Trenwith and since that time nothing further has been done at either mine. The lodes in Trenwith proved to be very small, but the quantity of water to be handled was singularly little and it was held by a small air pump.

It is said that the granite in which these lodes had been so very productive for tin further to the west in St Ives Consols makes its appearance in the lower levels of Trenwith and it is significant that some very good tin ore was broken in the lower levels of the latter mine during the last working. Indeed the operating company hoped to have developed Trenwith in depth for tin, but the disorganization and shortage of labour brought about by the last war completely upset all their plans and ultimately terminated the company's existence.

Section 3: St Just and St Ives Districts

Wheal Racer

Westward of St Ives Consols the same group of lodes has been extensively worked in the picturesquely-situated Rosewall Hill and Ransome United mines, where the level of the country rises to over 700 ft. above sea-level. On the western slopes of Rosewall Hill a small mine known as Wheal Racer has been sunk in the granite to a depth of about 60 fathoms and although the property has hitherto been worked with a conspicuous lack of energy it is considered locally to be a most promising prospect. The mine contains two lodes, one dipping north and the other south, and it is thought that they will junction at about 10 fathoms below the present bottom workings. These veins range up to 3 ft. in width and are very easy to break and in every way cheap to mine. Although there is no adit and the water is now standing nearly at surface the amount to be pumped is negligible and when the mine was last worked about 35 years ago there was insufficient water for the small mill during the summer months.

Having thus dealt with the more important and interesting mines in the northern group it is possible to describe briefly the outstanding mines in the southern portion of the district.

The Providence Mines and Wheal Margery

These celebrated old mines, situated on the killas-granite contact at the extreme north-eastern end of the Land's End granite mass, where it dips beneath the killas and greenstone, were worked for many years under the now flourishing holiday resort of Carbis Bay. The lodes were extensively explored seawards as well as under the land and were worked in both the killas-greenstone rocks and in the granite. In the former they were mainly productive of copper, but in the granite large quantities of tin were mined. As at Trenwith much pitchblende was found with the copper. Wheal Margery was worked on a single small lode to a depth of 165 fathoms below adit level and is recorded as having produced 16,400 tons of copper ore and 100 tons of black tin.

Section 3: St Just and St Ives Districts

The Providence Mines were formed by the consolidation of several smaller "setts" or leases in 1832 and they continued in operation until 1878. The records of output are somewhat confusing and obviously incomplete, but it is apparent that at least 15,000 tons of copper ore and considerably more that 10,000 tons of black tin were produced. There are numerous small but rich lodes in these mines, but there are two principal ones which intersect one another almost at right angles — the Laity lode, striking E. 20° N., and the Comfort lode, N. 15° W. At a depth of 105 fathoms from surface an important carbona was discovered about 80 ft. south of the intersection of the two lodes. The workings ultimately reached a depth of 200 fathoms from surface or 150 fathoms below adit and on a called-up capital of £11,569 paid 88 dividends totalling £111,020.

It is recorded that during the closing years of the old company's existence the bulk of the tin was obtained from the shallower levels and in view of the fact that the lodes have been extensively explored in depth in the granite there is good reason to suppose that the mine has, in fact, exhausted the productive tin zone. It is difficult, therefore, to see what justification existed for the re-opening of the property in 1907 and it is regrettable that so much capital was then uselessly expended in surface plant and in partially unwatering the workings. Down to the depth to which the mine was drained the lodes proved to be too small or poor to work and even if the whole property had been drained to the bottom it is very doubtful whether any better discoveries would have been made. As emphasized in the second part of Section 2 when an old mine, such as Providence, affords the clearest indications of having been worked until it has exhausted the tin zone it is better left severely alone in the future.

Giew Mine and Wheal Reeth

One and a half miles south-west of Carbis Bay the granite rises into the prominent eminence of Trink Hill, the highest point of which is 695 ft. above sea-level. Around the south-western slopes of the hill are to be found the remains of the Wheal Reeth and Reeth Consols mines. Incidentally these properties should not be confused with the Wheal Reeth in the Breage district that has been reworked within recent years.

Section 3: St Just and St Ives Districts

The Wheal Reeth on the southern side of Trink Hill is an exceedingly old mine, where it is recorded that an early "fire" or pumping engine was working as long ago as 1748. The mine was finally closed in 1867, when she had reached a depth of 250 fathoms. The records of production are incomplete, but at the time of the abandonment it was stated that £458,000 worth of tin had been sold and of this amount £78,000 had been paid in dividends. This was an exceedingly dry mine, the water often being insufficient for ore-dressing purposes.

Reeth Consols lies about a quarter of a mile north of old Wheal Reeth and has been worked from the deep valley west of Trink almost to below the summit of the hill. The returns of mineral from these mines are confusing, as various sections of the property have been worked at different periods under such names as Billia and Durlo, Reeth Consols, South Providence, South Wheal Speed and, finally, as the Giew mine. Under the latter title the mines were re-opened about 1909 by the St Ives Consoldiated group and as such continued in operation until 1922, when they were compelled to close in consequence of the great fall in the price of tin and the abnormally high cost of supplies resulting from the last war.

As the Giew mine the property was sunk during the last working to the 217-fathom level below adit and several extensive drives, of which the 142-fathom was the longest, were driven eastward under Trink Hill. Latterly the workings were nearly all to the east and, as these were all on one lode and had extended far beyond Frank's shaft — the most easterly shaft in the mine — ventilation was very unsatisfactory. Furthermore Frank's, which was the principal shaft, was an exceedingly bad one, being both inclined and very crooked. Consequently it was long ago proposed that a new perpendicular shaft should be sunk on the eastern side of the hill and an attempt was made to refloat the mine about 1928 with a view to putting this proposal into effect. Unfortunately, however, the scheme did not come to fruition and the plant has long since been removed.

Providing that a new shaft were sunk, thus permitting of adequate ventilation and a general reduction of working costs, there seems good reason to think that the mine would prove to be a sound speculation. The lode worked by the late company is a small one, rarely more than 3 ft. in width, but it is of fairly good average value. On the whole it is an easy and cheap one to mine, although there are hard "bars" in it and wherever these occur the lode becomes poor. It was one of these

Section 3: St Just and St Ives Districts

bars about 100 ft. long at the 142-fathom level that the old men had encountered when they abandoned the mine in 1877. The records of stope samples in the possession of the late underground manager show values up to 150 lb. of black tin per ton and at least one sample of 56 lb. over a width as great as 6 ft. The smallest width sampled in these records is 1 ft. and the deepest level in the mine shows values up to about 46 lb. per ton. The late company undoubtedly made a cardinal mistake in attempting to mill too large a tonnage; the mass of very poor ore that was sent to the mill lowered the grade to an entirely uneconomic level. Giew is typical of the smaller type of Cornish mine which only possesses a modest tonnage of ore and yet ore that is of a fairly good grade. When working such a mine it is folly to pollute the good ore with a large tonnage of all but worthless material merely in order to reduce costs per ton mined. Rather should success be sought by close attention to detail and economy in every department, including, as far as possible, all administrative and other overhead charges.

In the opinion of men who worked in Giew the lode split at the 142-fathom level and the northern or foot-wall portion has not been seen eastward below that point except in some stopes near Frank's shaft. Furthermore, as mentioned previously, there is said to be at least one other lode in the property that was not touched at all in the last working and it would seem therefore that very considerable opportunities yet exist in Giew for making further important discoveries.

Locally the mine is considered to be a very good prospect and given the necessary new shaft it is, in all probability, the best remaining proposition in the whole of this particular area. It is not without significance that at a point nearly half a mile east of the summit of Trink Hill and in a direct line with the Giew lodes there exist some shallow workings that are reputed to contain favourable propsects for tin. As long ago as the '60s of the last century it was proposed that the unworked tract of ground to the east of Trink Hill be explored as it was thought to contain great possibilities. The results obtained in the most recent working of Giew have greatly strengthened the reasons for that opinion and it is to be hoped that this most promising prospect will be given a thorough test in the not-distant future.

Section 3: St Just and St Ives Districts

Wheal Sisters

About two-thirds of a mile south of Giew and situated in the depression between the Trink and Trencrom Hills are to be seen the extensive dumps and ruins of the group of mines known as Wheal Sisters. The remains of the surface works, which extend along the strike of the lodes for over a mile, bear silent witness to the activity that prevailed here until the closing years of the last century. The underground workings are very extensive and numerous lodes have been worked in the granite, their average bearing being E. 25° N.

Wheal Sisters was formed in 1877 by the amalgamation of Wheal Margaret, Wheal Mary, Wheal Kitty (Polpeor), and Trencrom. Most of these small mines had had long and individually prosperous careers before the amalgamation took place. As in the case of nearly all the very old mines the records of production are seriously defective, but from the available returns it is apparent that the group sold at least 16,400 tons of black tin and 5,500 tons of copper ore and there is reason to believe that the total output greatly exceeded these figures. In 1907 the "Cornish Consolidated" group proposed to re-open these properties, but little was done beyond constructing an elaborate stone-built power station. An old miner who had worked for many years in these mines once remarked: "If anyone wants tin in Wheal Sisters the best place to go to is as far away from the mine as he can go!" In all probability this somewhat "Irish" expression of opinion sums up the prospects in Wheal Sisters very accurately. Like many another old mine that has been long and extensively worked for tin the ground has been so thoroughly explored that there now remains only the most slender hopes of being able to make any new discoveries of importance and there is, therefore, little justification for any further attempts to re-work the mine.

Summary

In this review of the two most westerly mining districts of Cornwall brief mention has been made of the more important and outstanding features

Section 3: St Just and St Ives Districts

of each area and of the principal mines that have been worked in them. It will be realized that both districts have been very extensively explored and there now seem to be but few opportunities there of making any fresh discoveries. Furthermore the existing mines as a whole cannot but be regarded as being definitely exhausted. Paradoxically, however, the few legitimate speculations yet remaining in these districts are amongst the best in Cornwall. Although the same cannot be said of the mines in the St Ives district that have received favourable mention it is strongly to be hoped that Trenwith and Giew will be given a further and more competently conducted trial in the future.

Section 3: The Marazion District

Section 3: Part II

THE MARAZION DISTRICT

In the first part of this section, the principal mines of the westernmost mining districts of St Just and St Ives were described at some length. It is now proposed to deal with the considerably larger though somewhat scattered mineralized areas that occupy the "instep" of the foot of Cornwall.

If reference is made to the key map on page 50 it will be noticed that these mining areas are tabulated as the Marazion, Wheal Vor, St Erth, Gwinear, and Crowan districts. The Wheal Vor area, incidentally, is more frequently referred to as the Breage mining district. Although it is convenient to divide this part of Cornwall into the several areas named, the lines of demarcation are entirely arbitrary, as almost the whole of this part of the county is mineralized to some extent from one coast to the other and each mining district merges gradually into another.

Apart from the relatively small granite outcrop of the Tregonning and Godolphin Hills in the southern part of the area and the much smaller exposure in the celebrated St Michael's Mount, near Marazion, the whole of the land surface is composed of sedimentary "killas" and interbedded greenstones interspersed with numerous intrusive "elvan" dykes. Nevertheless, as shown by the map, the area under review is bounded on the east and west by two of the County's major granite masses and there seems little reason to doubt that the granite underlies the whole of this part of Cornwall at no very great depth. It is questionable, however, whether the igneous rock would be reached by mining at economic depths throughout the greater part of the area, especially in the northern section. It would therefore seem that any further search for tin ores should be confined in the first place to those portions of the various mining districts that are reasonably close to the granite outcrops.

Commencing about half a mile west of the town of Marazion and continuing eastward along the shores of Mount's Bay for about $4\frac{1}{2}$ miles, the Marazion mining district, as commonly defined, extends inland for rather over a mile at its western end and for a distance varying from 2

Section 3: The Marazion District

to 3 miles at its eastern border, where it is bounded at its north-eastern extremity by the Hayle River.

With the exception of the small outcrop of granite in the island of St Michael's Mount the whole area is composed of killas, although the great granite mass of The Land's End approaches to within a short distance of its western borders and the much smaller granite intrusion of the Tregonning and Godolphin Hills bounds it on the east. As seen from the relatively high ground east of Marazion the view of the district is most interesting, the bold granite hills to the west and east rising far above the average level of the land, which throughout the greater part of the area varies from less than 50 to little more than 200 ft. above sea-level. When surveying the district from this vantage point near Marazion it is impossible to escape the conclusion that the granite probably underlies the whole area at no very great depth.

As so far developed this has been primarily a copper-producing district, although considerable quantities of tin have been obtained from several of the mines. The lodes are frequently most productive of copper when in contact with the elvans and are considerably wider than those of the majority of the other Cornish mining districts — Henwood estimated the Marazion lodes' average width to be 4.68 ft. The principal veins cross one another in a rather confusing manner, although on examination it can be seen that they are mostly confined to two general bearings — namely, W.N.W. and W.S.W. All the mines in the area are relatively shallow and it is doubtful whether their depth would average as much as 500 ft.

Many of the mines, especially those at the western end, near Marazion, are very old, several of them in their early days having been exceedingly profitable. Over a century ago a speculator (Thomas Saunders Cave) attempted to re-work many of them on an extensive scale. Unfortunately the direction of affairs was placed in the hands of the wrong type of man and there was much extravagant expenditure, so that in spite of a considerable production of mineral the great venture ended disastrously for Cave with a loss of £193,000 odd on 27 mines. Very few leases in the western part of the area have been seriously re-worked since that time with the exception of Tregurtha Downs, where several attempts at re-opening have failed, principally because the scale of operations has never been commensurate with the cost of pumping the great volume of water that has to be handled there.

Section 3: The Marazion District

East and West Wheal Darlington

Commencing in the Marazion Marshes, at the south-western corner of the district, the first mines to be noted are the Bog Mines — later known at East and West Wheal Darlington. The land under which they were worked being marshy ground and but little above sea-level they were very wet mines and when in operation a century ago the maximum quantity of water pumped exceeded 1,200 gallons per minute. The water coming from a cross-course was very salt, but that from the lodes was much less so.

There are two lodes in these properties and they are thought to cross one another in the intervening unexplored ground between the two mines, where one lode is also believed to fault the other. The North Lode of the western mine, having a bearing of 20° N. of W., is the South Lode of the eastern mine. Likewise it is thought that the South Lode of the western mine, which has a bearing of 10° S. of W., may be the North Lode in the eastern mine, as this, too, has the same bearing, but at the time Henwood was writing in 1843 this point had not been definitely established. Both lodes were unusually wide, especially the South Lode in the eastern mine where Henwood recorded widths of from 2 to 72 ft. Both veins dipped south and both yielded tin and copper ores, the recorded sales (probably very incomplete) showing an output of 18,900 tons of copper ore and 416 tons of black tin. The mines were worked to below the 110-fathom level and continued in operation until about 1857.

Wheal Crab (later Wheal Chippendale) and Wheal Virgin

Immediately east of the Wheal Darlington mines and worked under the steeply-rising ground north of Marazion are the very old properties known as Wheal Crab and Wheal Virgin. At one period these formed the western end of a large combination of leases known as "The Marazion Mines". Here there are two or three south underlying lodes, having a bearing of a few degrees S. of E., which have been worked in Wheal Virgin to a depth of at least 120 fathoms. The major produce of these mines

Section 3: The Marazion District

was copper ore, of which there are recorded sales in excess of 23,000 tons. Locally Wheal Virgin is considered to be a very good prospect for tin.

Wheal Rodney, Wheal Hampton, Tregurtha Downs and Owen Vean

Immediately east of Wheal Virgin and under the flat low-lying land north-west of Goldsithney village the lodes bear more towards the south, thus assuming a strike of E. 20° S., and have there been worked from west to east in Wheal Rodney, Wheal Hampton, Tregurtha Downs, and Owen Vean respectively. Although operated on a fairly extensive scale and at several different periods these mines are all relatively quite shallow and the writer is strongly of the opinion that they offer as good a prospect for tin on a large scale as any group of properties that can be found in the western part of Cornwall.

Wheal Rodney (or, more correctly, East Wheal Rodney), Tregurtha Downs, and Owen Vean are very old mines, but Wheal Hampton, which is situated between the first two properties, is a much younger venture that was opened up during the present century. Tregurtha Downs was reopened about 1882 and between that date and 1897 operations were three or four times suspended and resumed. A considerable quantity of tin was produced and at times a profit was realized, but the tremendous fall in the price of tin during the '90s, combined with the very heavy pumping costs and incompetence and mistakes in management and control, brought the whole venture to an inevitable end.

The quantity of water to be handled per minute in Tregurtha Downs varied from about 1,000 gallons in summer to a maximum of over 1,500 gallons in winter and this fact has hitherto been regarded as a fatal obstacle to any possibility of successfully re-working the mine. However, when pumping at this rate the neighbouring mine to the east — Owen Vean — was partially unwatered and it is fairly certain that considerable quantities of water were being drained from the other mines of the group to the west and Penberthy Crofts mine to the north. Therefore, instead of making the limited tonnage of ore obtainable from Tregurtha Downs carry the whole of this excessively heavy overhead charge, it is obvious

Section 3: The Marazion District

that the entire group should be jointly worked and on a very much larger scale than has hitherto been attempted. In view of the fact that the other mines mentioned do not seem to be unduly heavily watered the result of working them all together would probably be to reduce pumping charges per ton (on the much larger output then available) to a reasonable and manageable figure. At a later date it is hoped to discuss this question of larger-scale working as applied to the Cornish mines, but although the writer is of the opinion that the advantages are frequently more than outweighed by the disadvantages he is very strongly of the opinion that Tregurtha Downs and the neighbouring properties are a classical example of mines that must be worked on an altogether larger scale if success is to be achieved.

In passing it is not without historical interest that the very fine 80-in. cylinder Cornish pumping engine that handled such great quantities of water at Tregurtha Downs was later re-erected at the celebrated South Crofty mine and is today still going strong, in spite of the fact that she has worked on four mines and is now 89 years old! When at Tregurtha Downs this engine, which has a shaft stroke of 10 ft., operated 20-in. diameter plunger pumps and often ran in winter at the record speed of 13 strokes per minute. On one occasion the whole of the woodwork of the engine house was destroyed by fire, but within 10 days the repairs were sufficiently far advanced to enable the engine to be restarted and within about a month the whole mine was once again "in fork".

There are three principal lodes in Tregurtha Downs, all of which dip to the south, as well as a stanniferous elvan, which has yielded some very rich tin ore at times and which has been stoped out to a great width in the workings known as "the quarry". From north to south the ore-bodies are named the North, Middle, and South lodes respectively. The North lode, which is a blue one and about 3 ft. wide, is the richest in the mine. It has been worked extensively down to the 95-fathom level, below which a large winze has been sunk on it to the 110 fathom. The workings on this lode extend both east and west of the "Engine" or pumping shaft, although mostly to the east. Down to the 50-fathom level the lode is practically exhausted. The Middle lode is 3 to 4 ft. wide and has not been found below the 40 fathom. The South lode is a very red one, 6 to 8 ft. wide and of lower grade than the other two; it carries about 1% of recoverable black tin. The South lode has a flatter dip than the other two and, in consequence, the hanging-wall requires more support

Section 3: The Marazion District

to prevent it collapsing. The deepest point reached on this lode is the 110-fathom level, although not much has been done below the 95 fathom.

The water encountered in sinking the main pumping shaft was not at all heavy; the large flow was derived mostly from the lodes and especially from the stanniferous elvan wherever it was penetrated by the cross-cuts. The lodes, as is so often the case, were very porous and the bulk of the water followed the developments downwards so that latterly the upper workings above the 50 became relatively dry. The mine was operated by means of a perpendicular, though rather small, pumping shaft which was sunk below the 110-fathom level and, in addition, by at least six hoisting shafts, most of which were inclined and crooked. The adit is about 10 fathoms from surface and there were levels at the following depths (in fathoms):- 20, 30, 40, 50, 67, 80, 95, an intermediate level at the 100, and, finally, another main drive at the 110.

In an ill-advised attempt to offset the then heavy fall in the price of tin it was decided to economize in pumping by abandoning the lower half of the mine and to allow the water to rise to the 50. At this time some of the miners were of the opinion that the North lode had been lost below the 67-fathom level. They thought that the narrow vein then being worked in the deeper part of the mine was merely a branch and that the mineralization noticeable in the foot-wall of these lower workings was an indicator of another ore-body standing further to the north. Shortly before the decision to abandon the lower half of the mine was put into effect a few men working in a stope at the 95 fathom east of Waterstile shaft decided to prove the matter by cross-cutting in the wall of the stope. Within 3 ft. they encountered an exceedingly rich "cindery" lode 2 to 3 ft. wide and this was quickly followed by a similar discovery at the 100-fathom intermediate level. The mine captain and miners alike were filled with consternation when it was decided to proceed with the abandonment of these lower levels, but the decision having been taken it was apparently too late to obtain a reconsideration of the matter. The men who had made this discovery were given the opportunity of breaking as much ore on tribute as they could obtain in the few days remaining before pumping from the bottom levels ceased and in consequence they earned a considerable sum of money. Soon after the water had been allowed to rise to the 50- fathom level it was realized that there were no ore reserves above that point capable of sustaining the mine during a period of exceptionally low metal prices and in consequence operations

Section 3: The Marazion District

were entirely suspended. At a later period a further attempt was made to rework the mine, but the venture seems to have been ill-directed and lacking in capital and after the water had been only partially drained everything was stopped and the mine finally abandoned. The only records of production available are those covering the period 1886 to 1897, when 1,195 tons of black tin was produced, but it is obvious that the total output, including that from previous workings, must considerably exceed that figure.

In the early years of the present century work was commenced further west on a lode that is thought to be a continuation of one of those of Tregurtha Downs. The young mine thus brought into being was named Wheal Hampton and although it had a very chequered career, owing to the operating company being perpetually short of capital, it developed into one of the most promising prospects to be found anywhere in the western half of Cornwall. Only one lode has been worked and this, somewhat red, about 3 to 4 ft. wide (occasionally much wider), dips south at about 45°. There is a good deal of doubt as to which of the Tregurtha lodes this is, as it is not as red as the South lode of the latter mine and it appears to be standing too far to the north to be that ore-body. On the other hand although neither the colour nor dip of the Hampton lode is suggestive of the Tregurtha North lode it has yielded about 30 lb. of black tin per ton, which is much nearer to the average value of the North than the South lodes as so far worked in Tregurtha Downs.

Wheal Hampton was worked to a depth of 40 fathoms from surface by means of a single perpendicular shaft which intersects the lode above the 40-fathom level. The mine is connected to the Tregurtha adit, which comes into the property at about 10 fathoms from surface. Above the adit the lode was much split up, but from that depth down to the bottom level the lode gave good values and has been almost entirely stoped away for a length of over 400 ft. The writer has been informed by the then underground manager at Wheal Hampton that approximately 300 ft. east of the shaft the lode split in its course towards Tregurtha Downs and thereafter the separate branches were never worth more than about 10 lb of black tin per ton and consequently did not pay to mine. Westward consistently good values were maintained as far as the levels were driven, but as the old workings of East Wheal Rodney were close at hand and a considerable quantity of water was being cut it was necessary to suspend these developments when they were only 120 ft. west of the shaft. As the

Section 3: The Marazion District

property was then endangered by the Tregurtha water to the east and that of Wheal Rodney to the west and as the local Inspector of Mines was not willing to permit further drivage in either direction unless the respective mines were unwatered it was decided to drain Wheal Rodney, thus making available the high-grade ore standing intact on the Wheal Hampton western boundary.

With a view to testing the quantity of "coming" water in Wheal Rodney a small steam pump was installed in one of the old shafts by means of which the water was quickly lowered to the first level — approximately 10 fathoms below adit. However, when working a century ago and drained to the bottom workings, the maximum quantity of water that had to be handled amounted to 200 gallons per minute and as the Wheal Hampton pump had reached its limit and was dealing with 340 gallons per minute, much of which was percolating from Wheal Rodney, it was decided that it would be advisable to erect a large Cornish pumping engine on the latter mine to enable both properties to be drained immediately to the 40-fathom level. Unfortunately, however, the company had exhausted its slender cash resources and before the project could be completed the mines were abandoned towards the end of 1914. In addition to the Wheal Hampton plant the dismantled machinery at Wheal Rodney awaiting re-erection remained rusting away on the ground, but it has now long since been scrapped. Today only the partly-completed pumping engine house remains as a reminder of an ill-fated and inadequately financed venture. In passing, it should be recorded that the depth of Wheal Rodney, which has been variously stated to be anything from 52 to 80 fathoms, proved to be about the former figure when plumbed during the late company's operations and this depth agreed with an old plan then in their possession. The only records of production from Wheal Rodney show an output of 6,850 tons of copper ore between 1824 and 1848.

The second working of Wheal Hampton, which thus came to an end in 1914, lasted about five years, during the whole of which time the company lacked the capital necessary to open out the property in a miner-like manner. Although all drilling was done by hand not more than about 40 men were employed underground, but on surface there were approximately 120 employees. The ore had to be transported to the old Tregurtha Downs mill, a third of a mile distant, and much of the plant and especially the various steam engines employed were completely

Section 3: The Marazion District

worn out and exceedingly wasteful of fuel. It can therefore be seen that no matter how favourable the prospects working costs were bound to be high and the mine never had a chance to become firmly established. A very eminent mine manager who inspected the property during the last working told the writer that he was much impressed by the strength and regularity of the walls of the lode and the appearance which it had of being likely to persist both in strike and in depth. Furthermore the underground manager, previously referred to, says that in his opinion the lode at the 40-fathom or bottom level looked as favourable for further sinking as it had ever done at any point in its development. He states, however, that by reason of the company having exhausted its cash resources the whole of the available ground was stoped out and nothing of value now remains above the 40 fathom with the exception of the good ground on the western boundary. Incidentally it should be noted that the ore-shoots in Wheal Hampton pitch westward, while those of Tregurtha Downs incline to the east. Wheal Hampton itself is an exceedingly short mine on the strike of the lode and unless it can be worked in conjunction with Wheal Rodney and preferably with Tregurtha Downs, too, it is not worthy of further consideration. Provided, however, that the whole group could be tackled as one extensive unit the prospects of success would be excellent. It must be remembered that only one lode has as yet been developed in Hampton and Rodney and, given sufficient cross-cutting, there is every probability that the remaining ore-bodies of Tregurtha and possibly others too would be intersected. Although, with the exception of Wheal Hampton, these are all old mines it may be said that their development, both laterally and in depth, has as yet hardly been commenced and the very considerable ore-bodies already discovered and partially worked justify a much more spirited and competently conducted trial some day.

About 15 years ago a proposal was made to re-work the whole of this group by sinking a large perpendicular shaft to an initial depth of 1,000 ft.; the shaft was to be situated further south than the existing ones in order to cut the predominantly south-dipping lodes of the area at considerably greater depths. Although nothing came of the scheme it certainly appears to be a sound proposition and it is to be hoped that it will be put into practice at some future date. As a short-term temporary policy it would probably be quite feasible to explore and work the Hampton lode considerably deeper by making use of the existing per-

Section 3: The Marazion District

pendicular shaft at that property, providing that a pump were installed of sufficient power to enable Wheal Rodney to be unwatered at the same time. Until the mines were actually connected it might be necessary to pump independently in the Rodney shaft, as the plans of such old mines are not reliable and holing from one mine to another under these circumstances would not be a pleasant undertaking. If the authorities had taken a more enlightened and far-sighted view of the national tin requirements after the invasion of Malaya these mines should have been one of the first propositions to receive attention, as this plan for re-opening a part of the group without having to deal with a great quantity of water could have been put into action relatively quickly and cheaply. However, in common with many other exceedingly good prospects in the County, these mines will presumably have to wait until peace has once more been established before there will be any opportunity of resuscitating them.

It will have been noted that although Owen Vean, the property immediately to the east of Tregurtha Downs, was included amongst the mines of the group little further mention has been made of it. There is, however, little information available about this very old mine, which was in operation as long ago as 1750 and which has since been worked at varying periods to shallow depths with Tregurtha Downs. In any vigorous re-working of the latter property and its associated mines Owen Vean would undoubtedly come into the picture, as it contains the same group of lodes. Locally it is thought that if the western part of the group were actively developed, the water of Wheal Virgin would be encountered and that it would therefore be advisable to include the latter property in any general scheme of re-working. Very probably, too, other mines in the district would ultimately have to be considered, either by reason of the water problems or because of the increasing scope of the underground developments. In fact the writer cannot but help thinking that a well-planned and adequately-financed resuscitation of the Rodney-Hampton-Tregurtha group might well ultimately lead to an even more widespread revival of mining in the Marazion district. However until these central and most promising mines have been thoroughly proved it would be folly to expend large sums of capital on the numerous other old mines in the neighbourhood. It is because of this possibility of the revival of the whole district finally hinging upon the success or otherwise of the East Rodney, Hampton, and Tregurtha mines that what would otherwise be a disproportionate amount of space has been devoted to their description.

Section 3: The Marazion District

In passing it should be stated that the foregoing information has been obtained from many sources, but the writer especially wishes to acknowledge his indebtedness for many of the facts to Mr A. Curnow, one-time underground manager of Wheal Hampton, and Mr S. Burgan, of Marazion, a miner of almost world-wide experience, who retains an intimate knowledge of Tregurtha Downs and Wheal Hampton.

Gwallon

This is a very old but quite shallow mine, situated to the north-west of East Wheal Rodney. The important adit draining all the mines from Rodney to Tregurtha Downs and others further to the east discharges into the narrow valley at the western end of the property. The adit is here 15 fathoms from surface and the mine has been sunk to a depth of 40 fathoms below adit. Most of what is now known about Gwallon has been recorded by Henwood. He states that the South lode has a bearing of 5° S. of W., dips steeply south, yields both tin and copper ores, and is unusually wide, varying from 4 to 12 ft. An elvan forms the north or foot-wall of the lode. This is one of the mines worked so unsuccessfully by Mr Cave over a century ago.

Wheal Bolton, Rospeath, and Great Wheal Fortune

About three-quarters of a mile north of Wheal Virgin and Wheal Rodney are situated the very old copper mines known as Wheal Bolton, Rospeath, and Great Wheal Fortune. Tin has been seen in the lode at Wheal Fortune, but the production from all these mines has been practically confined to copper ores. The records of production of these mines are very confusing, as they have several times been re-worked in conjunction with other properties in the neighbourhood and it is now impossible to ascertain what percentage of the very considerable output was obtained from each individual mine. Wheal Fortune was in operation in the early years of the 19th Century and is said to have yielded

Section 3: The Marazion District

a profit of £100,000. Together with Rospeath and Wheal Bolton it was re-worked by Mr Cave, but a heavy loss resulted on the working of each property.

Wheal Bolton, which is the most westerly mine of the three, has been worked to a depth of 87 fathoms; Rospeath, which occupies the central position, is sunk to the 57 fathom, and Wheal Fortune on the east is down to the 144 fathom. The bulk of the information now obtainable about these mines is derived from the writings of Henwood. All three properties are worked on the same lode, which splits in its course westward thus forming at least three lodes in Wheal Bolton. At Wheal Fortune the lode, which was crossed by an elvan, had an east to west strike, a variable dip both north and south, although mostly south, and varied in width from 3 to 8 ft. In Rospeath and Wheal Bolton the lode and its branches had a bearing of 5° S. of W. and the widths varied from 6 in. to 4 ft. The North and South lodes in Wheal Bolton dipped north and the Middle lode mostly south.

Other Mines in the Marazion District

The mines already described constitute but a small proportion of the total Marazion mining district. For more than two miles further east and from the coast inland to the Hayle River scattered mines are to be found studded all over the map. In the aggregate they have produced a large amount of copper and some have yielded appreciable quantities of tin, but as yet few have developed into large, deep, or "permanent" mines. Many of these properties are very old and the information now available about them is of the scantiest nature and as there are such a very large number of these mines it is only proposed to mention briefly the more important ones amongst them.

Penberthy Crofts Mine — Eastwards of Great Wheal Fortune the same lode that was worked in that mine has been developed in the Penberthy Crofts "sett". During the period 1818 to 1825 some 8,697 tons of copper ore and 60 tons of black tin were produced. In 1837-8 the mine was operated in conjunction with Wheal Fortune as one of Mr Cave's group.

Section 3: The Marazion District

Trevarthen Downs Mine — This property lies south-east of Great Wheal Fortune and is situated between Penberthy Crofts and Old Wheal Prosper; it is worked on the lodes of the latter property which strike E. 40° N.. Collins states that the mine is 60 or 65 fathoms deep and is said to have been once rich in tin and copper. He adds the very significant remark: "A large lode which came down on a great mass of white mundic (mispickel), which at that time had no value, and the mine was consequently abandoned." In view of the frequency with which large bodies of arsenical ores have been found to occur in Cornwall between the shallow copper and deep tin zones one cannot but help wondering what lies beneath this mispickel in Trevarthen Downs. There is an old mining saying in Cornwall "Mundic rides a good horse", referring to the tin usually found beneath the arsenical ores.

Wheal Prosper — A very old mine worked on the same lodes as the above and situated about a third of a mile north of Wheal Hampton. In 1837-8 it was worked by Mr Cave in conjunction with Great Wheal Fortune, Wheal Bolton, Rospeath, Owen Vean, Tregurtha Downs, Penberthy Crofts, and Wheal Friendship. At a later date it was included in the "Marazion Mines" group under the title "Prosper United" and the production figures are therefore completely mixed up with those of other mines. In 1832-49 the mine is known to have sold 14,600 tons of copper ore.

Wheal Friendship, Guskus, and Wheal Ann — A group of properties lying between St Hilary Church and Penberthy Crofts mine. The returns from all three are very much mixed up, as at various periods each mine has been worked with one of the others. Available figures show an output of approximately 27,000 tons of copper ore, 840 tons of black tin, and 200 tons of blende.

Halamanning, Retallack, and Croft Gothal — These very old and extensive mines lie nearly a mile east of St Hilary Church. Halamanning has been worked at different periods in conjunction with each of the other two properties. Halamanning and Retallack sold 9,550 tons of copper ore during the period 1832-6 and Halamanning and Croft Gothal 13,400 tons of copper ore in 1851-8. This last working, which ended in 1860, resulted in a loss of £100,000. In view of the fact that two 70-in. cylinder pumping engines were then employed it is apparent that the inflow of water was heavy, as Halamanning is said to have been worked only to a depth of 90 fathoms. Locally there is a tradition that Halamanning is a

Section 3: The Marazion District

very good prospect for tin. When writing of Retallack in 1843 Henwood reported the presence of one copper lode which had a strike of 22° S. of W. and varied in width from 4 in. to 2 ft. In 1831-6 Retallack sold 10,019 tons of copper ore.

Tindene — Situated at the north-eastern extremity of the district and in the valley of the Hayle River, where it bends around sharply to the north-east, is to be found the old Tindene mine. This is a small and shallow property that was abandoned at a time of rapidly-falling tin prices in 1892 and which is thought to be well worthy of a further trial at some future date. Some of the shafts are sunk near the bottom of the valley and in consequence of the mine being situated beneath low-lying ground the amount of water to be pumped is fairly considerable. The records of output for the years 1887-92 show a total of 167 tons of black tin and 12 tons of arsenical pyrites.

In the southern part of the Marazion district and mostly close to the coast there are several old mines that have produced a good deal of copper and a small amount of tin. Some of these were very profitable in their day and if deeper sinking should prove the existence of a tin zone below the copper deposits there may well yet be a great future before this part of the area. The majority of these mines were worked on lodes having a bearing of E. 30° S. The more important properties are as follows:

Tolvadden Mine — The dumps and shafts of this mine are situated on the seaward side of the main Marazion-Helston road, about half a mile east of the town of Marazion. The same rich lode that was worked further eastward in Wheal Neptune was discovered here. From 1857 to 1866 the mine produced 10,750 tons of copper ore and 12 tons of black tin. Charles Thomas, writing in 1867, stated that the property "yielded £65,000 worth of copper. The mine was sunk 110 fathoms and became exhausted of ore, although many persons are of opinion that further deposits may be discovered by extended explorations."

Wheal Neptune — Still further east the same lodes were at one period very profitably worked in Wheal Neptune. Collins states that this was an old mine when re-opened in 1810 and that it stopped working about 1838. Charles Thomas states that the mine "gave very large profits from shallow formations of an unusually rich quality; the workings were prosecuted to a depth of about 120 fathoms, but the principal profits were obtained from ore found much nearer to the surface. The value of

Section 3: The Marazion District

the ore sold is said to have been about £400,000 and the profit to have been £200,000." At a later date an attempt was made to re-open the property, but the venture failed through lack of capital. Thomas stated that there was still a great extent of undeveloped ground towards the eastern boundary. The only records of production show an output of 13,760 tons of copper ore during the period 1815-1823 and in 1838.

Trebarvah — Otherwise Wheal Castle; situated between Wheal Neptune and the coast and lying immediately east of Perranuthnoe. Locally the property is thought to be a good prospect for copper. Collins states that it is "a mine always well thought of, but which has not been very extensively worked". In 1852-74 it produced 3,450 tons of copper, 6 tons of black tin, 1,728 tons of iron ore, and some ores of lead and zinc.

Wheal Speedwell — An old copper mine just north of Prussia Cove. The 30-fathom level was driven beneath the sea on a copper lode. According to J.Y. Watson the profits before 1843 amounted to £60,000. In 1819-54 it sold 11,356 tons of copper ore and half a ton of black tin.

Great Western Mines — This group includes the following setts: Wheal Grylls, East Wheal Grylls, Great Wheal Grylls, St Aubyn and Grylls, Grylls Wheal Florence, West Wheal Grylls and Wheal Boxer. The mines are situated at Kenneggy Downs, on either side of the Marazion-Helston road. Peter Watson, the celebrated mine "adventurer" and broker, resuscitated these mines and worked them as a single enterprise in the middle of the last century. Charles Thomas states that some rich discoveries of tin were made from which splendid dividends were paid for a while, but on the whole the mines seem to have been a disappointment and a falling price for tin caused their suspension. The records of production are more than usually confusing and in any case are probably very incomplete. Between 1850 and 1886 sales are recorded of approximately 2,000 tons of black tin, 900 tons of copper ore, and 65 tons of arsenical pyrites.

At the eastern margin of the Marazion district there are several small tin mines — such as, Millpool, Wheal Lemon, and Leeds and St Aubyn — that were worked in the killas close to the granite mass of the Tregonning and Godolphin Hills. Wheal Lemon has always been considered to be a good prospect. At Leeds and St Aubyn the writer has been told that wolfram has been found in the dumps.

Section 3: The Marazion District

Wherry Mine

Before leaving the Marazion area and describing the mines of the Breage district brief mention must be made of the Wherry mine at Penzance — one of the most singular and enterprising ventures ever known in Cornwall.

A part of an elvan dyke visible at low tide on the foreshore opposite Penzance Promenade was long ago known to be impregnated with granular cassiterite and permeated by thin strings of black tin. An attempt was made to work this in about the year 1700, but when the excavations had been carried down only a few feet work had to be abandoned as the rocks were 720 ft. beyond high-water mark and only accessible at low tide in calm weather. In 1778 John Curtis, a miner of Breage, together with others, made a further and successful attempt to work this unusually situated deposit. The first step was the sinking of a pumping shaft and this occupied three summers, as the rock is covered about 10 months in the year and during the winter the surf is so great that any work is impossible. At spring tides the rock is covered to a depth of 19 ft. When the shaft had been sunk a few fathoms a narrow collar, only 2 ft.1in. square and strongly built of wood and iron, was carried up to a height of 20 ft. above the rock and made watertight by pitch and oakum. A staging was built around the top of the collar and by means of hand-operated pumps it was then possible to continue working at all states of the tide. Considerable percolation occurred, however, as the mine was worked to within 3 ft. of the sea and the heavy swell and surf frequently made it impossible to remove the broken tin-stone from the rock to the beach. An engine was therefore erected in 1792 on the "green" above the beach and a wooden bridge constructed from the shore to the rocks, along which a line of "flat rods" from the engine to the shaft was supported. By these means it was possible not only to pump by steam power but they were enabled to convey materials and ore to and from the shaft at any state of tide and weather. As operated in this manner ore to the amount of at least £70,000 was produced and the venture was only brought to a close in about the year 1818 when a large American vessel broke from her anchorage and demolished the staging and machinery. A further attempt to work the mine was made about 1859, but in spite of a large expenditure it was not found to be

A. Wheal Kitty, St Agnes. 1920s

B. Tolgus Shaft. 1920s

C. Cligga Mine. 1920s

D. Polhigey Mine. Late 1920s

E. Wheal Agar. c.1912

F. Tregurtha Downs. 1890s

G. Geevor Mine. Victory Shaft. 1920s

H. Levant Mine. c.1900

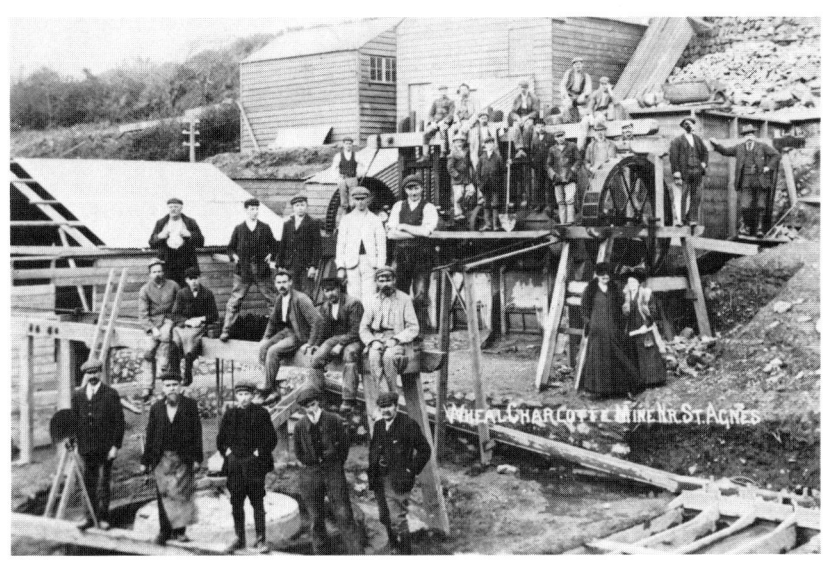

I. East Wheal Charlotte (not Charlotte). 1908

J. Prince of Wales Mine. Early 1900s

K. Botallack Mine. Allen's Shaft. c.1910

L. Wheal Peevor. c.1914

M. Wheal Vor. c.1908

N. Lambriggan Mine c.1900

O. Levant Mine. c.1900

P. Wheal Grenville. Fortescue Shaft. c.1910

Section 3: The Marazion District

profitable and was soon abandoned.

The workings were confined to the elvan, which was 18 ft. wide with a strike of E. 45° N., the dip being to the north. Hawkins states that "besides the small veins of tin which ran through this rock its whole mass was impregnated with tin to such a degree as to be worth the expense of raising, 15 ft. of the 18 producing 16 cwt. of white tin (about 24 cwt. of black tin) in 1,000 sacks (100 tons) and another foot as much as 1 cwt. of white tin ($1\frac{1}{2}$ cwt. of black tin) in one sack (2 cwt.)." Collins adds that "the ore though rich was very foul for it contained sulphuret of tin (stannite), sulphuret of copper and many other substances so that it was found necessary to 'burn' it. Indeed, this is said to have been the first mine where the ores were 'burnt' before dressing." Several tons of cobalt ore were obtained and Davy records having seen in the waste dumps the following minerals: Blende, oxide of uranium, oxide of titanium and of iron, pitchblende, nickel and arsenical pyrites.

Section 3: Wheal Vor or Breage District

Section 3: Part III

THE WHEAL VOR OR BREAGE DISTRICT

This comprises the greater part of the granite boss of the Tregonning and Godolphin Hills and the highly-mineralized and almost horizontal belt of killas which extends eastward for about $2\frac{1}{2}$ miles to the western slopes of the great Carnmenellis granite mass. The only lodes of major importance occurring in the granite are those of the Great Work group lying in the hollow between the Tregonning and Godolphin Hills. The majority of the other important ore-bodies are found in the killas immediately to the east of these hills and extending for not more than $1\frac{1}{2}$ miles from the granite. Including the lodes of Great Work in the latter rock the district is not more than about $2\frac{1}{2}$ miles wide from west to east and 3 miles long from north to south.

This area has been primarily a tin-producing district, notwithstanding the fact that the majority of the mines have been worked entirely in killas country rock. Although relatively small in area this district has produced some of the richest mines ever worked in the West of England and it would seem that there are still a few excellent prospects there awaiting further development.

Wheal Vor

Wheal Vor is one of the most important and possibly the richest tin mine ever worked in Cornwall. Practically the whole of its mineral riches occurred in the killas; where the lodes entered the granite of Tregonning Hill on the west they split up into strings and became almost valueless.

The commencement of operations here dates back to the 17th Century and it was either at Wheal Vor or the neighbouring Godolphin mine that gunpowder was first used for blasting in about 1689; the man who introduced the explosive to Cornwall was Thomas Epsley, of Somerset.

Section 3: Wheal Vor or Breage District

According to Joseph Carne the first steam engine in Cornwall was employed at Wheal Vor in pumping. Presumably this was a Savery engine, for what is believed to have been Newcomen's first engine in Cornwall was erected at these mines before 1714. This machine enabled the mine to be sunk to a depth of 60 fathoms below adit, but this working of the mine soon came to an end.

About 1812 the property was re-started by a Captain John Gundry and the first steam stamping-mill in Cornwall was erected there by the celebrated engineer Woolf. Captain Mark Reed employed forges underground to economize the miners' time in getting their tools sharpened, the ground being exceptionally hard. By 1826 the mine had erected its own smelting works and in the following year seven pumping engines were in operation, the maximum quantity of water pumped being over 1,100 gallons per minute. An appreciable quantity of copper was produced in addition to the very large output of tin, but in consequence of legal troubles the enterprise was brought to a close in about 1828. Operations were resumed in 1836 and by the following year were again very extensive; about 1,200 persons were employed and it was said that the mine had the appearance of a town. The output for some while amounted to 200 tons of black tin per month — more than one-third of the whole production of the County.

There were at least four lodes in this wonderful mine, but only two — the Main and Wheal Sosen — were worked to any extent. The Main lode has a bearing of E. 28° N. and a northerly underlie, varying from 10° to 25°. The Wheal Sosen lode lies further to the south and bears E. 24° N., but by reason of its much flatter underlie, which varies from 20° to 36° N., it junctions in depth with the Main lode. The latter has been worked for a total length of 3,900 ft. and to a depth of nearly 2,000 ft. from surface. It is intersected by at least three cross-courses, the western one of which — known as the Valley Flucan — occurs in about the middle of the property. The Flucan divides the lode into two great ore-shoots, both of which pitch eastward at about 45° away from the granite. The western shoot between the granite and the Valley Flucan was worked to a depth of 145 fathoms. The lode was rather small, usually varying in width from a few inches to $1\frac{1}{2}$ ft., although occasionally as much as 8 ft. East of the Flucan the lode opened out into an enormous orebody of unparalleled richness, the deepest workings on which extended to a depth of 295 fathoms below adit. Two elvans occur in this part

Section 3: Wheal Vor or Breage District

of the property and the lode was very productive in them. The actual vein varied in width from 3 to 30 ft., but the highly-mineralized walls of the lode were sometimes stoped out for a width of 50 ft. and in some instances to more than double that extent.

In spite of the immense tin output further litigation seems to have impoverished the owners, so that development was neglected and the lode becoming poor in the bottom the mine was stopped in 1847, although the surface works were carried on until 1850. In about 1853 the property was again re-opened and of the two pumping engines erected to drain the mine one had a cylinder 100 in. in diameter, with a stroke of 11 ft. — the first of this great size in the world. In this re-opening Wheal Vor proper formed only a portion of a very large group of mines that were to be worked as one immense united concern under the title of the "Great Wheal Vor United Mines". After many blunders and mechanical breakdowns the mine was ultimately unwatered to the bottom, but the lode there appears to have been found poor and hard and as the company was by that time involved in financial difficulties the mine was soon abandoned without doing any new development worthy of the least mention.

Collins in his celebrated "Observations" gives his reasons for thinking that further great discoveries would have been made had the deeper developments been carried further eastward, but Hunt thought that the great ore-shoot had been "unbottomed". Charles Thomas, commenting on this disastrous attempt to re-work Wheal Vor, remarked that "this eastern ground appears to have received but little trial, attention having been directed almost exclusively to the bottom of the mine". He concludes: "There never was any mining enterprise launched upon such a large scale which was so dreadfully mismanaged as this attempt to re-work the old mine, and never was there such a wanton waste of capital, £250,000 having been completely thrown away."

The story of depressing blunders in connnexion with the great old mine was, however, not yet ended, for in 1906 a further and even more disastrous attempt was made to resuscitate Wheal Vor. On this occasion electricity was employed for pumping, power being generated by steam on the property. A good deal of trouble was experienced with the plant and there were several changes of machinery. Ultimately, when the water had been lowered about half way, the breakage of a single bolt resulted in the complete wreckage of the only generating engine then in

Section 3: Wheal Vor or Breage District

operation. With extraordinary lack of foresight the company had neglected to provide for any such contingency and in consequence before the damage could be made good the mine was again completely flooded. As the company had expended so much capital and was unwilling to go any further in the matter the mine was once more abandoned and thus ended the last and most ill-conceived attempt to rework Wheal Vor.

When examining the plans and sections of the mine the writer has been struck by the notable absence of any extensive cross-cutting in the deeper workings. At Woolf's shaft a cross-cut has been driven south at the 106-fathom level for about 400 ft. and the Sosen lode intersected at that point. Likewise at the 115-fathom level another cross-cut has been extended 800 ft. north to intersect Trueman's lode. With these exceptions, however, there does not seem to have been any cross-cutting of any extent whatsoever in the deeper parts of the mine. The plans are strongly suggestive of that oft-repeated and fatal error in days gone by — undue concentration on one highly-productive lode without making an adequate effort to find additional ore-bodies to supplement or replace it before it has become impoverished.

Whatever the prospects are of making additional discoveries further eastward and in depth on the Main lode it seems to the writer that in any possible re-working attention should be first devoted to making a thorough exploration of all the ground between Wheal Vor and Wheal Metal, which lies a third of a mile to the south, and after that has been completed the country should be cross-cut as far north as can be economically carried out from the old mine. Immensely productive as Wheal Vor and Wheal Metal have been, the country between and to the north and south of these mines is entirely unexplored in depth and it may well be that parallel to these known ore-bodies vast and quite unsuspected riches still exist. Nevertheless, although there are still great possibilities in Wheal Vor, the writer would not advise any further attempt at re-opening the mine when there are so many equally promising and far more shallow and easily proved prospects awaiting development elsewhere in the County.

Section 3: Wheal Vor or Breage District

Wheal Metal

Of the several mines that were worked by the Great Wheal Vor United mining company one only proved a success, this being an old but hitherto little-developed property named Wheal Metal (known locally as Poldown). This developed into another extraordinarily rich mine and had the early profits derived from it not been swallowed up in the great losses incurred in old Wheal Vor and other mines of the group it might well have proved to be one of the greatest prizes ever discovered in Cornwall. When the Wheal Vor United company was ultimately forced to abandon all its mines, with the exception of Wheal Metal, it carried out a financial reconstruction in the hope of salving something from the wreck and continued to work Wheal Metal under the same title of "Great Wheal Vor United". This working of Wheal Metal continued until 1877, but as the output during this period appears under the heading "Great Wheal Vor" it has often led to considerable confusion with the returns from Wheal Vor proper.

There are at least four lodes in the property, but, as in Wheal Vor, only two have been worked to any extent — namely, the "Metal" lode and Schneider's. The former lode is the principal one in the mine; it bears about 26° N. of E., underlies to the north about one in three, and varies in width from 6 in. to 5 ft., although occasionally much more. Schneider's lode is a small one, hitherto only seen in the eastern part of the mine. Its outcrop is south of the Metal lode and it has a more northerly bearing, but as its underlie is much greater it junctions downward with the Metal lode, the point of junction declining from east to west. A large amount of tin was found at this point. The other lodes in the property are known as Vansittart's and the South lode, but very little has been done on them.

The Metal and Schneider's lodes proved to be wonderfully rich, but the values were far from being consistent. At times the lodes were almost barren of tin for considerable distances and then a junction with a secondary lode or a cross-course would produce an enormous enrichment, bands of nearly pure tin oxide 2 to 4 ft. wide being met with and continuing for many fathoms in length. Furthermore, as in Wheal Vor, the walls of the lode were often richly impregnated with tin oxide in the joints and laminations of the killas and "floors" or bands of impregnated

Section 3: Wheal Vor or Breage District

rock were found at times extending as much as 20 to 30 ft. from the lode. The *Mining and Smelting Magazine* for November, 1864, stated that:

> The lode in Ivey's shaft, sinking below the 184 fathom, turned out, for the last 3 ft. sunk, 6 tons 4 cwt. of black tin which even at present prices shows a money value of £800 per fathom... such a tin mine as this has not been seen in Cornwall during the present generation.

The *Mining Journal* for July, 1865, remarked:

> Probably no such course of tin was ever seen as that now opening at Wheal Metal. It has been proved 130 fathoms long by 30 fathoms deep, to average over £50 per fathom in value and in places £400 to £800. The aggregate value of 14 "ends" is over £1,700. Truly this is a wonderful mine — probably the richest tin mine in the world.

From 1855 until the end of 1858 the monthly output of black tin averaged 40 tons and during the years 1865-9 about 60 tons per month. When Charles Thomas was writing in 1867 the monthly returns were from 70-80 tons.

The mine continued to yield largely until the early '70s, but throughout its whole career it was worked with the most astounding prodigality and lack of adequate development. It appears as if the reconstructed Wheal Vor company was determined to obtain the utmost possible profit from Wheal Metal for the least possible expenditure, so as to recover as much as possible of the capital previously lost in the other mines of the group. On more than one occasion the mine was nearly abandoned when it had become temporarily poor and the developments had been shamefully neglected. Each time, however, a further discovery was made, sometimes merely by chance, so that the life of the mine was once more prolonged. In 1874, however, when a depth of about 220 fathoms had been reached, the mine had once more become poor and the shareholders who had bought in at high prices while large dividends were being paid were unwilling to put up the capital required for carrying out explorations that could no longer be deferred. Two eminent mine managers were called in to advise, but, in spite of their recommendation that the western ground be explored and cross-cutting be carried out north from Edwards' shaft, little was done and soon after the mine was abandoned — to the intense disgust of the local miners. Operations on a small scale were continued for a while at an old shaft further west, but in 1877 the company was wound up and the plant sold. Thus ended the ill-starred

Section 3: Wheal Vor or Breage District

and reckless working of one of the richest tin mines ever known in the West of England. Collins states that during a period of 20 years less than £25,000 was expended on the mine, although within 14 consecutive years £95,000 was paid out in dividends, equal to 22% of the tin sales during that period. He gives the output from 1854 to 1877 as being 9,190 tons of black tin.

As in the case of old Wheal Vor there can be little doubt that Wheal Metal is as yet only partially explored and it may be that further great discoveries will be made there at some future date. Whatever the prospects may be of the existing lodes yielding further discoveries in depth or further westward it is obvious that the mine yet possesses very great possibilities, for as shown by the plans practically no cross-cutting whatsoever has been done in depth. Indeed the mine is an even more striking example of this cardinal error in development than old Wheal Vor itself. Collins expressed the view that there was no mine in the West of England deeper than 100 fathoms that more deserved to be reworked than Wheal Metal. While the writer cannot say that he by any means agrees with that opinion, there is little doubt that the prospects in Wheal Metal are probably far better than those in the neighbouring Wheal Vor.

Great Work

A mile to the north-west of Wheal Vor and seated in the depression in the ridge connecting the Tregonning and Godolphin Hills is the exceedingly old and very famous Great Work group of mines. Numerous rather narrow lodes with a general bearing of E. 45° N. have been worked here from time immemorial, the whole of which were situated in the granite with the exception of the extreme eastern workings on the "Great Work" lode which are in killas down to about the 100-fathom level. A well-known though very peculiar fact in connexion with the mines of this area is that the riches of Wheal Vor were almost entirely confined to the killas, the lodes becoming very poor in the granite, whereas in Great Work precisely the opposite state of affairs prevailed and lodes that had been exceedingly rich in granite were found to be practically valueless in the killas.

Section 3: Wheal Vor or Breage District

The eastern workings on the Great Work lode in Great Work proper were ultimately extended to a depth of 185 fathoms below adit, or 203 fathoms from surface, which was considerably deeper than any other section of the mine. It appears, however, that the bottom of the property was abandoned shortly after the middle of the last century and although the mine was worked almost continuously until 1902 the scale of operations and the depth from which the ore was mined became progressively less. Indeed the property gave every indication of having been unbottomed and largely exhausted by the "old men" and it is difficult to see what justification could have possibly existed for re-opening the mines a few years ago. The major shoot of ore on the Great Work lode is produced by the junction of that vein and the Wheal Reeth lode a short distance east of the Leed's shaft. East of that point the two lodes continue as one ore-body to the eastern end of the mine, the ore-shoot declining in depth eastward so that it lies parallel to the killas-granite junction. When unwatered in recent years practically the whole of the ground east of the Leed's shaft proved to have been stoped out. The bottom of the mine at the 185-fathom level below adit at Deerpark shaft was ultimately reached, but apart from cutting a few samples no work whatsoever was done and the water was again allowed to rise in these lower workings. Meanwhile attention was directed to the development in shallow ground of the two lodes west of the Leed's shaft. A considerable amount of new development and stoping was carried out in this part of the property in addition to the stripping and sampling of old levels, but on the whole the results were very disappointing. The values were comparatively low and as the lodes were narrow it was not an economic proposition to work them and when a short-lived depression in the tin markets intervened in 1938 the property was once more abandoned.

A further section of the mine previously worked on the Wheal Breage lode was very ill-advisedly re-opened at about the same time as an independent concern, but here, too, the results were as might have been expected; practically everything of value had been exhausted (with the exception of some ground in the western part near Barker's shaft). Furthermore the Wheal Breage company did not even succeed in completing the unwatering of their part of the mines and they closed down after spending a considerable amount of capital without doing any further development worthy of mention.

Wheal Reeth — the third section of the property that has been re-

Section 3: Wheal Vor or Breage District

worked in recent years — consists of comparatively shallow workings at the extreme western end of the Great Work lodes. These numerous ore-bodies are usually very narrow, although the values are sometimes high. However, unless the price of tin is maintained at a fairly high and steady level, it is doubtful whether the working of these lodes will ever be an economic proposition.

It cannot be too frequently stressed that, when a mine like Great Work has been extensively worked for tin over a long period of years and where the records show every sign of a gradual exhaustion of the ore-bodies, it is plain folly to expend a large amount of capital there in the hope of finding further good values. By contrast a mine such as Wheal Metal that has been recklessly and irresponsibly worked and then abandoned immediately it ceased to be profitable is in an entirely different category and may yet merit further development.

In passing it should be recorded that the only figures of production for Great Work show that between the years 1832 and 1902 some 6,256 tons of black tin and 1,020 tons of copper ore were sold. The total production, however, must have very greatly exceeded these figures, for the mine was being actively worked long before the end of the 18th Century.

Great Wheal Fortune

Immediately south of Wheal Metal, and worked in the same belt of killas that has been so productive there and in old Wheal Vor, is a mine named Great Wheal Fortune. This property must not be confused with the mine of the same name in the Marazion district.

The Great Wheal Fortune, in Breage, has been worked on the very flat south-dipping Carnmeal lode and upon a rather extraordinary series of tin-bearing veinlets known as the "Conqueror", "Copper Ore", "Blue Burrow" and "Elizabeth" branches respectively. These mineralized belts course about E.N.E. and dip steeply to the N.N.W. They are intersected and considerably faulted by an elvan 40 ft. thick, which courses a few degrees south of east. The mine is notable for the large open quarries worked on the outcrop of the Conqueror branches. Collins estimated the

Section 3: Wheal Vor or Breage District

yield of ground quarried to have been about 9 lb. of black tin per ton of stuff. The Carnmeal lode was very rich at times and was ultimately worked to a depth of 150 fathoms below adit before the mine closed down in 1868. The only records of production, covering both the mine and open quarries, show an output of approximately 3,000 tons of black tin and 300 tons of copper ore.

Wheal Trewavas

On the coast, near Trewavas Head, two and a half miles S.S.W. of Wheal Vor, is the isolated Wheal Trewavas copper mine, which was worked in the extreme southern limits of the Tregonning Hill granite mass. The ruined engine houses of this very picturesquely situated mine remind one of a lesser Botallack.

Henwood states that there were four lodes, although he only mentions one as carrying copper values. Collins remarks that the mine was worked to a depth of 70 fathoms on several lodes. The returns show that durin the period 1836-46 17,385 tons of copper ore were sold. According to Symons, writing in 1884, the mine was holed to the sea and could not therefore be re-worked.

Godolphin Mine

North of Wheal Vor there are several lesser mines — such as, Penhale Wheal Vor, Polladras Downs, Polrose and the Breage Mine (Gwyn and Singer), some of which have been worked at different periods with Old Wheal Vor. Hitherto none of these properties has been of major importance, but they are nearly all relatively shallow and there still seems to be plenty of scope for further development, especially in depth.

A mile and a half N.N.W. of Wheal Vor and situated at the northern foot of the Godolphin Hill is the exceedingly old and important Godolphin mine. In the writer's opinion the geological situation of the

Section 3: Wheal Vor or Breage District

property, combined with the records and traditions of its past history, strongly suggest it as being one of the best speculations for tin that can be found in any part of Cornwall. The property, which has been very rich in copper ores to a depth of 120 fathoms, has been worked in the killas at the foot of a granite hill and it seems a reasonable assumption that the granite exists not far below the present bottom workings. From the writings of Henwood and others it is apparent that tin was present in nearly every one of the several lodes and tradition states that very rich tin ores were making their appearance in the lower levels when the mine was abandoned by a typically "copper" management in about 1847. In view of the fact that the mine is situated in one of the richest tin districts ever known in Cornwall and has every appearance of having reached the transition stage from copper ores in killas to tin in granite it seems to be an exceedingly good prospect. The chief drawback to the property is the very large amount of water that has to be pumped, but most of this is probably surface water, as the mine is situated beneath the low-lying swampy valley of the Hayle River. As the latter is little more than a large stream at this point it would probably be quite feasible greatly to reduce this percolation by constructing a concrete channel for the river bed and carrying out other effective drainage works throughout the length of the property.

Collins states that this is one of the oldest mines in Cornwall, it being an important producer as far back as 1678. Large profits are known to have been made there before the 19th Century. The only records of production, which are obviously fragmentary, cover various periods between 1815 and 1847 and these show an output of 9,803 tons of copper ore and 10 tons of black tin. Henwood records several lodes in the mine varying from 1 to 6 ft. in width and having a general south-easterly bearing with the exception of one which strikes 25° N. of E. When writing of the mines of this district in 1867 Charles Thomas stated that "the Godolphin Mine was found very rich in copper close to the surface, and large profits were derived from it.... The explorations were continued in more recent times, by the aid of very powerful steam machinery below the points at which the lodes were found to produce copper, and large quantities of tin were raised; but the low price of that metal, and the heavy expense of draining the mines, prevented their being profitably continued."

Finally, before taking leave of this district, the writer would like to

Section 3: Wheal Vor or Breage District

draw especial attention to the large area on the north-eastern slopes of the Godolphin Hill which has never yet been systematically explored, although it has long been thought to be an exceedingly promising speculation. The Godolphin Hill rises over 400 ft. above the valley in which the old Godolphin mine is situated and were a deep adit driven southward into the hill from the valley a very large piece of ground could be explored without incurring any pumping charges. The West Godolphin mine, that was worked on the north-western shoulder of the hill, produced 1,500 tons of black tin during the years 1870-90 and, as many of the lodes of this mine pass eastward into the area under consideration, there seems all the more reason for giving this piece of ground a thorough trial. Hitherto the only explorations carried out in the north-eastern part of the Godolphin Hill have consisted of small adits and shallow surface pits and these are said to have exposed several most promising small lodes.

Section 3: Part IV

THE ST ERTH, GWINEAR AND CROWAN DISTRICTS

When describing the Marazion and Wheal Vor sub-divisions of the mining area that lies in the "instep" of the foot of Cornwall it was pointed out that any division of this part of the county into separate districts is entirely arbitrary as the whole of the country from one coast to the other is more or less mineralized. Of this area it is true, however, to say that the mining districts of St Erth, Gwinear and Crowan constitute a single and indivisible unit and it is therefore proposed to deal with them as such in the present article.

The area under consideration forms approximately a rectangle, its length from east to west being some 6 miles and breadth from north to south rather less than 4 miles. Although there are a few small and unimportant mines near the north coast between Gwithian and Hayle, the northern boundary of the district may for all practical purposes be taken as the main railway line from Camborne to Hayle. The western boundary is formed by the Hayle River, which, from the point near Relubbus where it bends sharply eastward, also constitutes the southern limit of the area, thus dividing it from the Marazion and Wheal Vor districts. On the east the region is bounded by the granite hills of Crowan — the western fringe of the great Carnmenellis granite mass. Throughout the greater part of this area the low undulating surface is less than 300 ft. above sea-level and in the main it slopes towards the Hayle River on the south and west and the low muddy creeks of the Hayle estuary to the north-west.

With the exception of the granite which borders its eastern margin the whole of the area is composed of sedimentary "killas" rock with numerous intrusive dykes of elvan and interbedded greenstone. At its southern boundary, however, the district all but makes contact with the granite of Godolphin Hill, while the major Land's End granite mass outcrops within two miles of its north-western extremity. It is worthy of note that in several of the mines in the neighbourhood of Gwinear

Section 3: St Erth, Gwinear and Crowan

smooth pebbles of granite and even globular masses of that rock up to a foot in diameter were encountered in the highly-brecciated conglomerate which encloses many of the lodes, notwithstanding that on surface the nearest granite outcrop is several miles distant. Salmon states that in West Rosewarne mine boulders of granite of various dimensions up to 6 ft. were found in the lode, while a breccia and conglomerate of elvan, killas and granite also occurred.

The metamorphic aureole extends over a large proportion of the whole district and even in parts far remote from the granite outcrop there are isolated areas of slightly-altered sedimentary rocks, suggestive of an undulating floor of granite extending throughout the whole region. Notwithstanding these indications of the relative proximity of the igneous rock it has never yet been encountered in depth in massive form except in a few mines at the extreme eastern margin of the area on the killas-granite junction. Admittedly most of the mines are very shallow, only a few having reached a depth of a thousand feet, but whether deeper sinking would reach the granite in many of them at economic depths is entirely a matter of speculation. It is worth noting that the Geological Survey estimated the probable depth of the granite throughout the greater part of the area as being 4,000 to 5,000 ft.

Salmon, writing in 1864, remarked:-

It is essentially a copper district, that metal making in conjunction with elvans; and although it has been fairly productive, yet on the whole it has been economically the worst district in West Cornwall from the extremely bunchy and uncertain nature of its metallic deposits.

Unfortunately it must be added that subsequent explorations have only tended to confirm that opinion. Charles Thomas, nevertheless, writing in 1867, thought that there were distinct possibilities still existing there and even today it must be admitted that these extensive areas contain an immense amount of entirely unexplored ground. Furthermore the possibilities of mining for tin in the neighbourhood of the killas-granite junction at much greater depths than anything yet attempted in this part of Cornwall are entirely unexplored. Nevertheless, after all these possibilities have been taken into consideration, the district as a whole does not appear to offer many attractive prospects. If ever an infallible method of locating deep-seated ore-bodies is discovered this district, like other parts of Cornwall, may be found to contain much unsuspected mineral wealth and it is therefore wise to retain an open

Section 3: St Erth, Gwinear and Crowan

mind on the subject. Having said thus much, however, the writer has to admit that he has a strong prejudice against the region as a whole and as this series of articles is being written with a view to stimulating interest in the future possibilities of the County rather than in historical events it is not proposed to deal at any great length with the majority of the mines of the area.

Before mentioning briefly some of the more interesting and important properties in this part of the County there are a few further facts of general interest that should be noted. In general the lodes have a bearing of about W. 25° S., but in the centre of the area there are numerous "caunter" lodes striking approximately W. 30° N. and yet a futher series having a nearly east to west strike. The majority of the lodes dip south and their average width was estimated by Henwood to be 2.9 ft. The elvans, which abound, have widely varying directions, although, like the lodes, the majority of them strike W.S.W. There are numerous crosscourses having a general N.N.W. bearing, the most important of which is that which runs from the Alfred Consols mine in Gwinear to the east of Godolphin Cross and thence to Wheal Fortune, in the parish of Breage. This important cross-course shifts the whole of the lodes which it intersects, but the Geological Survey state that "its exact effect and the amount of its throw are difficult to ascertain for according to the mine plans its effects are very uncertain."

Great Wheal Alfred

Commencing in the north-western part of the area the first mine of note and, indeed, one of the most productive copper mines in the whole of these districts is the celebrated Wheal Alfred, which has been worked at times, in association with neighbouring leases, as Wheal Alfred Consols. These mines lie about three-quarters of a mile south of Angarrack and on the western side of the steep valley in which that village is situated.

The average width of the lodes here is 2.5 ft., but the large caunter lode, which is traceable eastward into the Herland mine, varies from 6 ft. to 25 ft. in width. Carne states that the lode dips north at an angle of 72°, traversing obliquely in its descent a large elvan dyke 300 ft. wide,

Section 3: St Erth, Gwinear and Crowan

which likewise dips to the north, although at a lesser angle. In the slate above the elvan the lode varies from 6 ft. to 9 ft. in width and only yielded a moderate quantity of ore, but on entering the elvan it became much richer and attained its maximum width of 25 ft. Below the elvan the width declined to 10 ft. and, the lode having become poor, it was abandoned.

As long ago as the year 1800 Wheal Alfred was being intensively worked, 1,500 people being employed at that time. Between 1804 and 1815 83,000 tons of copper ore were produced which gave a profit of £140,000. Shortly afterwards the mine was abandoned, but it was re-opened in 1824 and continued in operation until 1831; this working, it is said, resulted in a loss of £80,000. A further reworking, which commenced in 1846 and ended with the final closing of the mines in 1864, was stated by Collins to have been responsible for a further loss of £150,000. At about the same period the neighbouring mine to the east, Alfred Consols, was also being worked. Henwood states that it produced 42,000 tons of copper ore between 1844 and 1861, resulting in a profit of £98,000. Whether the two properties were worked separately or in conjunction it is now difficult to ascertain, as the returns of ore are very mixed and confusing statements have been made as to the nett financial results of this period of operation. Collins, referring to the working of Wheal Alfred Consols from about 1846 to 1864, states that at the latter date "the newer portion was drowned out owing to the flooding consequent on the stoppage of Great Wheal Alfred proper."

Charles Thomas, writing in 1867, was of the opinion that both mines were completely exhausted and one can only conclude that it was probably fortunate that the proposal made in 1907 to re-open them came to nothing. A small quantity of tin was sold by Alfred Consols, but the presence of tin in small quantities with the copper ores of this district is so widespread as to be of no particular significance. The apparent remoteness of these mines from the granite, the lack of any other particular reason for expecting that their lodes will become stanniferous with increasing depth, and the very large inflow of water to be handled combine to make any further reworking of these properties about as unattractive a speculation as anything that can be thought of in Cornwall.

Section 3: St Erth, Gwinear and Crowan

Mellanear

In the Alfred group there are several lesser mines, but the only other of importance is West Alfred Consols, which sold 8,400 tons of copper ore during the years 1851 to 1865. In 1864 the western part of the property was re-opened as the Mellanear mine and this continued in operation until 1889, the production during this period being 66,000 tons of copper ore and 80 tons of black tin. Mellanear is of interest as being the last of the great Cornish copper mines to be worked on an extensive scale exclusively for the red metal and also one of the last to be controlled by Messrs. John Taylor and Sons, who at one period were associated with many of the greatest West Country copper enterprises. The mine lies about half a mile south of Hayle, the shafts and dumps being situated on both sides of the main road running from Hayle to Praze and Leedstown. This was one of the wettest mines for its size ever worked in Cornwall, about 1,100 gallons per minute being pumped in winter time.

Mellanear contains two lodes, which are continuous with some of those worked in the Alfred Mines further eastward. Both of them have an E.N.E. course and underlie to the north, but only one appears to have been extensively worked in depth. The ore-body was fairly productive, although of low grade; the ore-shoot dipped westward. In 1876 the old unlimited liability or "cost book" company was succeeded by a limited one and notwithstanding the very depressed and continually-falling copper market the latter company had by 1884 paid dividends totalling more than its original capital. The mine continued to be productive for copper to below the 100-fathom level, but then became poor. A discovery of tin gave rise to the hope that the lode might become stanniferous with increasing depth, but, these hopes being falsified by later developments, the mine was abandoned in 1889.

There is a persistent local tradition that Mellanear is a good prospect for tin, but this is not borne out by any of the known facts. The writer has been told by the son of the late underground manager that the only tin of value was encountered at the 140-fathom level, west of the western pumping shaft. Several men worked this very profitably on tribute, thus giving rise to the tradition of good tin values in the bottom of the mine. At the time of this discovery the company did indeed entertain great hopes that tin would succeed the copper in depth and

Section 3: St Erth, Gwinear and Crowan

the western main shaft was accordingly sunk to below the 150-fathom level and another drive extended at a depth of 150 fathoms. However, at that horizon the lode proved to be hard and absolutely barren and although the developments were extended under the point where the tin had been discovered in the level above nothing whatsoever of value was seen. As no other encouraging prospects existed in the property the company decided to go into liquidation. The chairman's speech at the company's final meeting (a copy of which is in the writer's possession) makes it quite clear that the abandonment of the mine was due to three factors — namely, the falling off of copper values with increasing depth, the low price of copper, and, finally, because the hopes of discovering payable tin values beneath the copper had been completely falsified by later developments.

It can only be concluded that as has so often happened with other old mines "distance lends enchantment to the view" and mineral (in men's minds) continues to grow under water! Whatever may be said against any further attempt to rework Great Wheal Alfred seems to apply equally forcibly to Mellanear and it is to be hoped that further capital will not be wasted at some future date in proving once again what has already been demonstrated by a past generation.

Herland

Immediately east of the Alfred Mines and a little way south of Gwinear village is the very old and celebrated Herland mine. It is known that one of the Newcomen pumping engines worked here as long ago as 1746. Work below the adit level ceased in 1807, but operations were resumed in 1816 and the few statistics of production available show that between the latter date and 1843 18,517 tons of copper ore were produced, the profit from which is said to have amounted to £90,000. In addition to its principal product, copper, the mine yielded several other metals, the most important of which was silver, which occurred mainly as native and horn silver, pyrargyite, and silver glance. The silver production came principally from between the 90 and 120-fathom levels on a silver-bearing cross-course and to a lesser extent from one of the copper lodes. From

August, 1799, to April, 1800, the cross-course yielded over 115 tons of silver ore, which was partially smelted on the mine and yielded £5,649. Before 1814 over £8,000 worth of silver ores had been sold.

According to Henwood there are five lodes, varying in direction from E. 20° N. to E. 35° N., all of which dip steeply south, an important caunter lode striking 35° S. of E. and dipping S.W., numerous crosscourses having a bearing of a few degrees W. of N., and at least 4 elvans coursing about N.E. and S.W. The lodes are all very small, varying in width from an inch to two feet, although very occasionally flat veins or "floors" of quartz spreading out into the laminae of the slate on both sides of the lode were found to contain sufficient copper ore to be worth working for a distance of 10 ft. from the vein itself. In addition to being very small the lodes were often ill-defined and split up into branches and could not be depended upon to continue productive for any distance. In fact it was an exceedingly "bunchy" and patchy mine and the occurrence of ore was very sporadic and uncertain. In 1843 the workings had been carried down to a depth of 152 fathoms and extended laterally for a great distance.

Henwood recorded the presence in Herland at a depth of about 110 fathoms of nodular masses of fine-grained and decomposing granite entirely surrounded by slate, these masses varying in size from that of a hazel-nut up to 2 or 3 ft. in diameter. A still more unusual feature which he observed were the globular concretions in the Badger lode composed of "round masses of granite, slates and elvan, indiscriminately mixed and cemented together by a basis which is sometimes of felspar and quartz with a little mica, and at others of quartzose slate, iron, and copper pyrites."

Relistian

About a third of a mile east of Herland and situated near the village of Wall is the very old Relistian mine, known to have been worked as long ago as 1715. Here there are two lodes, the northern one of which strikes 20° N. of E., dips steeply north, varies from 1 to $3\frac{1}{2}$ ft. in width, and has been worked mainly for copper to a depth of more than 155 fathoms. The so-called South lode is in reality a large felspar-porphyry elvan, about 11 fathoms wide, some parts of which are quartzose, containing copper and iron pyrites and cassiterite irregularly distributed throughout the mass. This great elvan, which strikes 16° N. of E. and dips slightly S., has been entirely worked away in places for the cassiterite which it contained and in Henwood's day an open excavation on this dyke extended from surface to a depth of 39 fathoms, some of the workings being 30 ft. wide. The mine was noteworthy for the spheroidal masses of slate and quartz enclosed within the country slate and the numerous globular masses of felspar and elvan that were found in the lodes, together with globular concretions of rounded slate fragments cemented together by copper ore and cassiterite or even by a quartzose or slaty matrix. The only recorded sales of mineral show an output between 1832 and 1842 of 12,150 tons of copper ore and a few tons of tin after 1852.

Trevascus

Three-quarters of a mile N.E. of Relistian and a little way south of Gwinear Road station is to be found the Trevascus mine, which also dates back to the early years of the 18th Century. It was rich for copper as long ago as 1739, but when Henwood wrote in 1843 tin was the principal produce.

The mine contains two north-dipping lodes — the main one, striking 35° N. of E. and dipping at 70° to 86°, and the South, whose strike is 30° N. of E. and which has a dip of 80° to 84°. The main lode is very large, varying from a few feet up to 10 fathoms in width. Often there was no decided division between the lode and the killas and greenstone country in which it occurred. It seems to have been a mass of altered and mineralized rock, which could frequently be distinguished from the "country" only by the metallic minerals which it contained. It is still possible to descend a few feet below the surface in the wide outcrop workings on this lode, which was mined to a depth of more than 70 fathoms below adit. Henwood has recorded the presence here too of rounded masses of slate in the lodes and also globular lumps of copper pyrites surrounded by quartz. The only records of production, obviously

Section 3: St Erth, Gwinear and Crowan

fragmentary, show an ouput between 1838 and 1842 of 110 tons of black tin and 1,190 tons of copper ore.

Charles Thomas in 1867 thought the ground south of Trevascus worthy of further investigation, as hitherto it had only been tested by a few shallow adits and apparently that state of affairs still persists to the present day. Thomas also recommended the further development of Wheal Hartley, a large "sett" or lease lying east of Trevascus which has as yet only been developed on a small scale to a trifling depth.

Duffield

About two-thirds of a mile E.S.E. of Trevascus is to be found the site of the old Duffield mine, known at an earlier period as "The Weeth". The shafts are situated on either side of the road from Barripper to Carnhell Green. The mine is said to have been worked on a considerable scale and in 1842 had been sunk to a depth of 98 fathoms. The quantity of water pumped was heavy, amounting to 670 gallons per minute in June, 1841.

Henwood records the presence of five lodes having an E.N.E. bearing, two elvans with a N.E. strike, and two cross-courses and a slide. The latter faults all the lodes, but the cross-courses are peculiar inasmuch as they fault some of the lodes, yet are themselves heaved by some of the mineral- bearing veins. Henwood states that at the 88-fathom level on the South lode "in several places on both walls there is between the lode and the country a layer of slaty matter, containing many round stones. Some of these are elvan, others quartz, but the greater number are of slate. In some places this bed passes gradually into the slate; in others there is a well-marked and distinct junction." He also records the presence of tin in the deepest workings on this lode, but the only records of output show that between 1815 and 1824 and 1831 and 1841 7,800 tons of copper ore were produced. The last period of working which ended about 1849 is said to have resulted in a loss.

Section 3: St Erth, Gwinear and Crowan

The Rosewarne Mines

East of Relistian and south of Trevascus the numerous mines of the Rosewarne group are scattered over an area of approximately one mile square. The more important of these properties are Rosewarne United (formerly Gwinear Consols), New Rosewarne, East Rosewarne, Rosewarne Consols, Rosewarne and Herland United, West Rosewarne United and North Rosewarne.

With the possible exception of Rosewarne United it is doubtful whether the working of any of these mines resulted in a profit. Their major product was copper, although nearly all of them sold a little tin. Arsenic, blende and a little arsenical silver ore have also been produced in small quantities. In common with most of the other mines in the northern part of the Gwinear district the country rock in which their lodes occur is extraordinarily broken and disturbed and the lodes, which contain much brecciated material, were unusually "bunchy" in value. Very promising discoveries have been made in a number of these mines at various times, but the values rarely lasted for any considerable distance and nearly all these properties have been great disappointments. Admittedly very few of them have been worked to any appreciable depth — few much exceeded 100 fathoms — and it is obvious that there still remains a great deal of unexplored ground. Nevertheless the history of the district gives no reason to anticipate any better results in the future and apart from an entirely new discovery at surface one would not recommend further expenditure on any of the mines in this area.

Wheal Jennings (or Parbola) and South Rosewarne

These mines lie about a third of a mile S.E. of the village of Wall. Although not economically important they are interesting by reason of the unusual nature of their ore-body. This consisted of an east to west non-porphyritic elvan 40 to 80 ft. wide, which traverses a soft killas, the dip of the elvan being 40° – 60° S. The tin ore occurs in innumerable small parallel strings and shrinkage cracks running across the elvan from killas to killas at about 20° W. of S. These cracks varied in width from 4 in. or 8 in. to a size so small that 30 could be counted to the inch. In the neighbourhood of these small veins the elvan is completely altered and impregnated with ore. The cracks sometimes continued into the killas, but generally terminated at the wall of the elvan. Stopes have been worked yielding 10% of black tin, but ore-bodies of this type are notoriously "bunchy" and uncertain in value. A good deal of money has been expended on the mine again during the present century, but apparently without successful results. The only records of production show that at various widely separated periods of working a total of 358 tons of black tin has been sold.

Mines in the Neighbourhood of Leedstown

On either side of the road leading from Praze-an-Beeble to Hayle (*via* Leedstown) numerous old mines can be seen scattered all over the country for over a mile both east and west of Leedstown. The innumerable shafts and dumps bear witness to the extensive nature of the mining operations that were once carried on in this part of the county, although, somewhat strangely, very little information is now available about any of the mines concerned. It is known, however, that the whole of these properties were worked entirely in the killas and few have reached any considerable depth; copper was the major product, although tin has been raised in considerable quantities especially from some of the workings west of Leedstown. Although at surface the granite is approximately two miles distant to the south and east one cannot but help wondering whether the igneous rock does not exist much nearer to the surface than is generally thought probable. Little if anything is known that would warrant the re-opening and deeper sinking of any of these mines for tin and yet were some infallible method of detecting deep ore-bodies to be discovered it would not be altogether surprising to discover that the area still contains extensive and valuable ore-bodies that are as yet entirely unworked.

The statistics of production of many of these mines are extremely confusing and it is now very difficult to determine even approximately

Section 3: St Erth, Gwinear and Crowan

from which particular area a given output was obtained. Many of the leases appear to have been worked at varying periods as parts of different combinations and under different titles and in consequence the records of output are hopelessly mixed up. The following brief notes of the more important mines merely serve to give an approximate idea of their past importance and output.

Clowance Mine — At Howe Downs; from 1815 to 1824 sold 7,580 tons of copper ore.

Wheal Treasury — At Burnt Downs, east of Leedstown; from 1826 to 1844 sold 6,787 tons of copper ore.

West Treasury — Immediately west of the above; said to have been once rich. From 1845 to 1854 sold 9,500 tons of copper ore.

Wheal Tremayne (including *West Wheal Providence*) — Situated north and west of Leedstown. From 1848 to 1868 sold 5,260 tons of copper ore and 1,492 tons of black tin. Total production, especially of tin from West Providence; probably much exceeded these figures. Said to have paid considerable dividends from both tin and copper.

Lambo — Afterwards formed a part of Wheal Tremayne. From 1815 to 1824 sold 8,932 tons of copper ore.

To the south and west of Leedstown numerous scattered mines are to be found distributed over a tract of mainly agricultural land that is about 3 miles long from north-east to south-west and rather more than a mile in width. This area is bounded on its southern and western sides by the Hayle River and in view of the proximity of the granite at Godolphin Hill, immediately to the south, it would seem possible that that rock may exist at no great depth below some of the mines concerned. The more interesting and important mines in this area will now be briefly mentioned.

Carzise

This is an old mine, approximately half a mile W.S.W. of Leedstown, which has been repeatedly worked and abandoned. It was extensively developed a century ago to a depth of 85 fathoms on two south-dipping lodes. The main lode, according to Henwood, had a strike of 9° S. of

Section 3: St Erth, Gwinear and Crowan

W. and varied in width from 6 in. to 5 ft.; it was crossed by a caunter lode striking 12° N. of W. which was from 1 to $2\frac{1}{2}$ ft. wide. Both lodes yielded copper and tin ores, the latter apparently in considerable quantities. The only records of output show sales between 1820 and 1842 of 4,128 tons of copper ore and 185 tons of black tin. This was one of the properties in the Gwinear district that figured in an unpleasant promotion incident about 23 years ago. A certain amount of surface plant was erected, but nothing whatsoever was done underground and the mine was again abandoned.

Wheal Osborne

About a quarter of a mile west of Townshend is a small mine known as Wheal Osborne which, in view of its favourable situation, seems worthy of far more attention than it has received in the past. This hitherto unimportant property was opened up on the western continuation of some of the lodes that were worked in the old Godolphin mine. In both these properties and in the intervening unexplored ground between them the lodes run approximately parallel to the granite of Godolphin Hill; Wheal Osborne, moreover, lies in the same ore-parallel as the once very productive Great Work mine to the south and other, though lesser, tin mines to the north. Although the dip of the granite at surface does not make it seem probable that that rock will be reached in Wheal Osborne except at great depths it does not follow that this dip will be found to continue indefinitely in depth. Furthermore the history of Wheal Vor and other once exceedingly-productive tin mines in the neighbourhood suggests that, although the position of the ore-bodies has some relationship to the subterranean granite surface, yet, nevertheless, the great deposits of tin in this part of Cornwall are as likely to be present in the killas as they are to be found in the granite.

Whether these lodes at the northern foot of Godolphin Hill prove to be productive in the killas near to the granite, or in the latter rock itself, their development in depth is a thoroughly good speculation and one that may yield a rich return to anyone with sufficient foresight and energy to tackle the proposition boldly. It is not without interest that

Section 3: St Erth, Gwinear and Crowan

Wheal Osborne, which has sold small quantities of tin and copper, is said to contain a large "low-grade" tin lode (about 1%), which at the time it was last worked, about 70 years ago, was of far too low a grade to be thought of much value at that time.

Lewis Mine

This is a very old property, $1\frac{3}{4}$ miles W.S.W. of Leedstown. From 1852 to 1861 it produced some 1,275 tons of black tin and a large but unknown quantity at an earlier date. It was worked to a depth of 130 fathoms. The lodes here were frequently very narrow indeed, but the ore was so concentrated that it sometimes paid to drive a level on a vein little more than half an inch in width.

Wheal Gurlyn

This is another old mine, lying about a quarter of a mile south of the Lewis mine; it has been restarted on several occasions. It was last worked on a small scale and in an ineffective manner in the early years of the present century. It was later acquired by the interests who at that time were working Wheal Hampton, near Marazion, and who, as explained in an earlier article, were continually in financial difficulties and therefore unable to install the more powerful pumping plant then necessary at Gurlyn; consequently little further was done at the mine by its new owners. There are stated to be six tin lodes in the property, but during the last working operations were confined to one with a north-westerly strike which is thought to be a continuation of the West Godolphin caunter lode. This vein, which was only explored to a depth of 20 fathoms, was reported to average 2 ft. in width and to contain excellent values. This is a small and shallow mine which appears worthy of a much more extended and energetic trial.

Tregembo

This property is a small tin mine in the bend of the Hayle River, immediately east of Relubbus village. The whole of the tin obtained from the property is said to have been derived from one "bunch" of ore in a lode, where it "made up against an elvan". In spite of local statements to the contrary there appear to be no prospects that would warrant the re-opening of the mine. An early type of pneumatic stamp battery was employed here during the working in the '80s of the last century, when 112 tons of black tin were sold.

Returning to the centre of the area under review, some of the most important mines in the whole of this part of Cornwall are to be found in the tract of country south-east of Leedstown. From west to east the most important of these mines are Binner Downs and the Crenver and Abraham properties which together constitute an almost continuous series of workings for a distance of more than 2 miles east of Leedstown.

At the eastern end of this mineralized belt the granite of the Crowan hills was encountered in the Trenoweth mine and it is worthy of note that granite can also be seen in the dumps of South Crenver, approximately a quarter of a mile south of the eastern part of the Crenver and Abraham mines. Henwood has pointed out that the ore-shoots of Binner Downs incline towards the east, while those of Crenver and Abraham pitch westward.

If it be assumed that the ore-bodies dip away from the nearest mass of igneous rock (as almost invariably occurs in Cornwall) it is suggestive that in addition to the major granite outcrop at the eastern margin of the district there is also a subterranean boss of granite existing somewhere in the neighbourhood of Leedstown. The fact that the lodes of Wheal Tremayne and Carsize, north and west of Leedstown respectively, yielded tin in appreciable quantities may have some connexion with such a hidden dome of granite near Leedstown.

If indeed such granite does exist its importance in connexion with any future working of Binner Down cannot be overemphasized. Hitherto

Section 3: St Erth, Gwinear and Crowan

these mines have been worked solely for copper, but good tin values are reported to have been making their appearance in the bottom workings.

Binner Downs

This mine contains at least four lodes, one of which, the South lode, was worked to a depth of 185 fathoms, the other to lesser depths. The South lode, which is the main one of the mine and the same vein that was so extensively worked in the neighbouring Crenver and Abraham mines, bears 29° N. of E. and dips south at 52° — 78°. It varies in width from 10 in. to 8 ft. and has yielded mainly copper pyrites, although cassiterite and small quantities of other copper ores, as well as galena, blende, and iron pyrites, occur. The North lode, which strikes 25° N. of E., is nearly perpendicular; it varies in width from an inch to $1\frac{1}{2}$ ft. and contains mostly copper pyrites. There are also two caunter lodes — the Gooseberry (or Wheal Strawberry) lode, striking 25° S. of E., dipping steeply south and varying in width from 2 to 3 ft., and Hicks' lode, which strikes 17° S. of E. and intersects the North lode. The country rock is a deep-blue slate with a gentle N.W. dip. Henwood records the presence of a cross-course, a slide and a flucan, but none of these fault the lodes.

A large, although very variable, quantity of water had to be pumped at Binner Downs. According to Lean it varied during the years 1833 to 1837 from approximately 600 to over 2,000 gallons per minute. It is very doubtful, however, whether the actual quantities pumped ever equalled these figures, which merely represent the theoretical output of the old pumping engines.

It is known that the mine was worked very profitably during the first half of the 18th Century and it was again in operation during the years 1819 to 1838, when 51,056 tons of copper ore were sold. Collins states that the latter working resulted in a profit of £100,000. After the mine was abandoned £8,000 worth of tin was recovered from the dumps. There is a persistent local tradition that good tin values were encountered in the deepest workings and it is thought that the property is a good prospect for tin in depth. In view of the relatively shallow depth of the mine and the possibility that the granite may indeed exist

Section 3: St Erth, Gwinear and Crowan

at no very great depth in this part of the area it certainly seems that Binner Downs merits a further trial at some future date.

Crenver and Wheal Abraham

These celebrated and extensive mines, which also include Wheal Sarah and the old Oatfield mine, lie immediately south of the grounds of Clowance. They were amongst the major copper producers of the county. It is known that the Crenver and Oatfield portions were at work as long ago as 1801, at which date they were among the deepest and most extensive mines in the county, having then reached a depth of 140 fathoms below adit or 170 fathoms from surface. All records of early production are lost, but it is known that the mines were wonderfully rich in copper and it is stated that in 12 years 106,725 tons of copper ore was produced from one lode. By 1822 Wheal Abraham, having reached a depth of 240 fathoms, was the deepest mine in Cornwall. This working of the group terminated in about 1825, by which date it is said that a profit of £200,000 had been realized. With the exception of some small intermittent workings in outlying parts of the property which were carried on for a little while longer nothing further was done until the whole group was re-opened in 1864.

Before detailing the later history of these mines it should be mentioned that they contain at least four lodes, only one of which, however, has been worked or even explored below adit level. In fact these properties seem to have been a classic example of the vicious practice of rapidly exhausting a very rich ore-body without making any attempt to discover others to replace it. It was in connexion with these mines that, when asked why they had not done more exploratory lateral development, a one-time underground agent there made the famous remark: "We had a good mine and did not think of looking for another." Incidentally this policy has been as largely responsible as any other single factor for the premature decline of many of the great Cornish mines in the past. It is worth noting that it is the opposite practice of extensive cross-cutting and lateral development along known lodes that is largely responsible for the success during recent years of such modern Cornish mines as Geevor and South Crofty.

Section 3: St Erth, Gwinear and Crowan

The Crenver main lode is the same vein as that worked extensively further westward in Binner Downs. In the former property it has an east and west bearing and underlies south 15°. It is mostly $2\frac{1}{2}$ to 3 ft. wide, although at times apparently considerably larger in the lower levels. Carne states that it is often greatly mixed up with elvan, so that "it appears to be one component with the lode," which in such cases is generally poor. The South Crenver lode, which is a caunter, apparently met the Main Lode west of Henry Vivian's shaft and resulted in a great body of ore. The mine contains two elvans and several cross-courses. Of the three or four parallel lodes to the south it was said at the time of the re-opening in 1864 that two were sufficiently close to be worked from the existing shafts, but, strangely enough, even in this, the second and last working, no attempt seems to have been made to explore them.

The writer is unaware of any evidence that the granite was ever actually reached in these properties, but the killas-granite contact is seen immediately to the east in the Trenoweth mine, which was worked on the same series of lodes. Thomas states that the Crenver lode was very rich in copper near the contact, but that on meeting the granite, dipping westward under the killas, it split into branches and became altogether unproductive. It must be admitted that the granite of the Crowan hills has been singularly barren of mineral, but if at a point further westward this great lode had been explored considerably deeper and well below the granite surface it is an open question whether it might not have become a great tin producer in depth.

Subsequent to the abandonment of the mines in 1825 it was widely reported that tin had been encountered in considerable quantities in the lower workings. In consequence of these reports and the success then being achieved elsewhere in Cornwall by sinking for tin below the copper zone it was decided in 1864 to re-open the mines primarily with a view to working them for tin. This reworking, which lasted until 1876, was recklessly conducted and was a failure, resulting in a heavy loss of capital. After great expenditure the bottom workings were finally unwatered and a small amount of new sinking was done, the deepest shafts being extended about 20 fathoms to a final depth of 248 fathoms below adit.

Like Binner Downs these mines are heavily watered, three Cornish pumping engines being employed in the last working handling 1,000-1,400 gallons per minute. Much trouble was experienced with the pump-

Section 3: St Erth, Gwinear and Crowan

ing plant operating in bad crooked compound shafts and pumping difficulties are said to have been one of the main reasons why the mines were not more vigorously developed in depth. Some further considerable sales of copper ore amounting to 21,475 tons were made, in addition to which 151 tons of black tin and 751 tons of tin ore were sold. However, the venture proved a failure and following the death of one of the largest shareholders the mines were abandoned with a heavy loss in 1876.

It has been said that the tin which it was hoped would be discovered beneath the copper did not exist in worth-while quantities. At Dolcoath and other deep Cornish mines, however, it has frequently been demonstrated that the transition from copper to tin is gradual and there is sometimes an intermediate and more or less barren zone between the two ores. The plans of Crenver and Abraham, which the writer has examined, show so little new development in the bottom of the mine during the last working that it seems reasonable to assume that the possibilities of tin occurring in quantity in depth are as yet entirely unproved. Furthermore it is evident from these plans that even in the last working no attempt was made below the adit level to explore by crosscutting the several more or less parallel lodes that are known to exist. One cannot but help thinking that immense possibilities may still exist in these great mines, yet in view of the failure of the last working one would hesitate to recommend anyone to tackle them afresh unless further favourable evidence about their prospects in depth could be obtained by less expensive means than the unwatering of such an extensive series of workings. If ever the question of mining on an extensive scale in this area were to come under consideration again the writer would recommend that the neighbouring Binner Downs mine should first be unwatered and vigorously developed in depth. Such a programme would probably throw a most revealing light on the possibilities or otherwise of the considerably deeper and much larger Crenver and Abraham group.

In passing it is worth recording as a matter of historical interest some important events which occurred in connexion with the Crenver and Abraham mines. It was here, in about 1801, that the then manager, Captain Joel Lean, introduced the displacement plunger or "pole" which superseded the bucket type of pump. Shortly afterwards he made important improvements to the steam engines which resulted in drastic economies of fuel. The combined effects of his inventions and improvements was so to reduce the cost of pumping that whereas the mines had

Section 3: St Erth, Gwinear and Crowan

previously been worked at a heavy loss it was then possible to operate them very profitably to a much greater depth. At a later date the same engineer was responsible for the introduction of the practice of recording and publishing the "duty" (or work performed for a given quantity of fuel) of the engines at each mine. This practice produced an intense and healthy spirit of rivalry between the managements and engineers of the various mines throughout the county, who exerted the utmost efforts to improve the efficiency of their engines. In consequence a revolutionary reduction in the cost of power was achieved that alone made possible the era of comparatively deep mining that followed. It was estimated in 1835 that the annual saving to the Cornish mines resulting from the improvements initiated by Lean was of the order of 60%, or 100,000 tons of coal, which at that time had a monetary value of £80,000.

An event of an unusual nature in connexion with the Crenver and Abraham mines was the disaster that occurred on August 6, 1806. Heavy rain deluged the valley in which the adit discharges and the water flowing into the adit ran back into the mine and flooded it to a depth of 300 ft. in 15 minutes, drowning 7 men, while 50 others only escaped with difficulty. East Wheal Rose, the only other property in the history of Cornish mining to be drowned out by an inrush of surface water, was, peculiarly enough, like Crenver and Abraham, inundated by a cloudburst on a summer day.

Finally it is a matter of some historical engineering interest that a link with the last working of Crenver and Abraham still exists in the form of the 80-in. cylinder Cornish pumping engine at the Robinson's shaft of the South Crofty mine. This notable old engine, designed by the celebrated engineer Captain Samuel Grose, is one of the finest examples still in existence of the Cornish engine-builders' art. It was made at the "Copperhouse" foundry at Hayle in 1854 for the Alfred group of mines, previously described. After the abandonment of those properties the engine was removed to Crenver and Abraham at the time of the 1864 re-opening and there operated a set of pumps in the deepest shaft in the mines. After the closing of Crenver and Abraham she was removed to the Tregurtha Downs mine, near Marazion, and later still re-erected at her present site at South Crofty in 1903. In her 89th year this notable old machine is now pumping from the greatest depth in her career — namely, 2,020 ft. There are several unusual features in the design of this engine, the most important of which is the early attempt to apply

Section 3: St Erth, Gwinear and Crowan

the principles of uniflow steam expansion. Should the time arrive when the engine is no longer required by her present owners she should most certainly be preserved by one of the engineering societies as a memorial to the great Cornish pioneers who did so much to make deep mining possible.

Mines South of Binner Downs and Crenver and Abraham

About a third of a mile south of the Binner Downs and Crenver and Abraham groups there are a number of secondary copper mines which, collectively, have produced a considerable tonnage of ore, although, in comparison with the mines already described, they are of little importance. These lesser mines were worked on a series of lodes having a general south-easterly bearing, some of which in their course north-westward intersected or junctioned with the lodes of the Binner Downs and Crenver Mines.

Wheal Strawberry. — The most westerly of these smaller mines is Wheal Strawberry, also known as Wheal Julia. This is situated south of Binner Downs and was worked on a lode which is recognized as being the "Gooseberry" Lode of the latter mine. In Wheal Strawberry the vein varies in width from 6 in. to 3 ft. down to the 79-fathom level, its bearing being 25° S. of E. and the dip 56° to 86° S. The recorded production of copper ores during the years 1830 to 1837 amounted to 9,716 tons. Over 400 gallons of water per minute were pumped when the mine was working in 1837.

Wheal Dumpling. — The next mine east of Wheat Strawberry and on the same series of lodes is Wheal Dumpling, which is also known as Crowan Consols. Little information is now available about this old property; from 1863 to 1866 it sold 1,000 tons of copper ore.

Wheal Courtis or Drym. — Still further east and lying a full third of a mile south of Crenver and Abraham is Wheal Courtis, known also as Drym. The property is said to contain five lodes, although when in operation a century ago work was confined to one of them. In its early days it is said that the mine was very productive of tin and copper ores

Section 3: St Erth, Gwinear and Crowan

and gave a profit of £150,000. Reworkings in the '40s and '60s of the last century, however, were apparently failures.

South Crenver. — North-east of Wheal Courtis are to be found the remains of the old South Crenver mine — originally worked by the Crenver and Abraham proprietors and later reworked by independent companies during the '50s and '60s of the last century. During the last working 5,270 tons of copper ore were sold. Granite can be seen in the old dumps of the mine; it seems reasonable therefore to presume that the main body of granite of the Crowan Hills immediately to the eastward was encountered in depth. Incidentally the property has only been sunk to a depth of approximately 800 ft. from surface.

Having thus briefly reviewed all the mines of any importance in the Crowan area it remains to be pointed out that the granite hills forming the eastern margin of the Crowan and Wheal Vor districts themselves contain several small, shallow, and hitherto unsuccessful tin mines. What is probably the most important of these small scattered properties is Polcrebo Downs, just over half a mile due east of South Crenver. Although good discoveries were reported to have been made here at various times nothing of a permanent nature was found and the only recorded sales of black tin amount to 102 tons during the period 1884 to 1890.

In 1867 Charles Thomas remarked: "These granite hills of Crowan have been singularly barren of metals, not one profitable mine having been met with in them either for copper or tin." Since Thomas's time several attempts have been made to work lodes in these hills, but without exception the enterprises have been entirely unsuccessful and at the present time the writer is unaware of a single prospect existing in these hills that seems to warrant further expenditure.

Conclusion

The extensive area described in this article has in the past yielded several very important deposits of copper and although tin in smaller quantities is widely distributed throughout the region no major discoveries of that mineral have yet been made. Admittedly, with but few exceptions, the mines have only been worked to very modest depths, but although

Section 3: St Erth, Gwinear and Crowan

considerably deeper sinking might prove the existence of a widespread tin zone beneath the copper the indications are that it is predominantly a copper-producing area, which nevertheless, contains a certain amount of tin interspersed with its copper ores.

New methods of prospecting for deep-seated ores may one day throw an entirely different light upon the possibilities of this area, but at present it can only be said that very few of the mines offer any prospects that merit the expenditure of much new capital. The few mines that do appear to warrant further investigation have each been noted and in this connexion the writer would like once again to draw particular attention to the possibilities of the but little developed lodes at the northern foot of Godolphin Hill.

Section 3: The Wendron District

Section 3: Part V

THE WENDRON DISTRICT

If reference is made to the key map on page 50 it will be seen that the sixth district indicated on the map is that of Wendron, which is situated in the Carnmenellis granite area. This extensive tract of granite is one of the five major igneous intrusions of the West of England and at its northern fringe are to be found the most important mining areas of Cornwall. The present article, however, is confined to the small Wendron district that occurs in the south-western part of the granite mass.

The area may be briefly defined as follows: Commencing near the margin of the granite at Lower Town, one mile due north of Helston, a straight line should be drawn north-eastward to Porkellis Bridge, a distance of $3\frac{1}{2}$ miles, after which a further line is drawn almost due north for $1\frac{1}{2}$ miles, terminating at the southern slopes of Carnmenellis Hill. This constitutes the western boundary of the area, its eastern limit being defined by a parallel line drawn 2 miles to the eastward of that already described. It will thus be seen that the region is about 5 miles long by 2 miles wide and so covers an area of approximately 10 square miles.

With the exception of the small strip of sedimentary or "killas" rock at its extreme south-western margin the district under review consists entirely of granite interspersed with a few elvan dykes. The lodes, which are mainly tin-bearing, have a general W.S.W. strike and are, on the whole, rather narrow. With but few exceptions they have not been found productive at greater depths than about 500 ft. from surface. The exceptions are confined to the southern part of the area nearest to the killas, where, in all probability, denudation of the granite is much less than it is further north, nearer the centre of the mass, and even here only two mines are known to have reached a depth of 1,200 ft. Taking the district as a whole it is apparent that denudation of the granite has been very extensive and in consequence only the lower portions of the lodes now remain intact. Some of the debris of the upper parts of these ore-bodies is to be found in the alluvials of Porkellis Moor, which have been worked for tin from time immemorial.

Section 3: The Wendron District

The shallowness of the Wendron mines has often been emphasized as a point in their favour, but in view of the aforementioned facts there seems little prospect of reviving mining in the locality by indulging in deeper sinking. Furthermore the chances of being able to discover new lodes at shallow depths seem very slender indeed, for denudation has probably exposed them all and every indication of mineralization that outcropped was thoroughly investigated by the "old men". Until the second half of the last century this part of the County was largely wild moorland and under these conditions outcrops are easily discovered. An example of unsuspected exploration on the part of the old miners occurred in about 1928, when a part of the East Lovell mine was being re-opened. During the course of surface excavations a productive and very promising looking lode was discovered in a field where there were no signs of any mining operations having taken place. It was decided to sink a trial shaft on the lode, but within a few feet of sinking an old men's stope was encountered and it was soon apparent that the whole of the lode had been extensively mined! With the exception of an area south of Porkellis village the writer does not know of any prospects in the Wendron District that warrant any further expenditure. As in any old mining field there is always the possibility of a chance discovery being made, but on the whole the prospects are not encouraging.

In the whole of this district there are only five or six mines with a *recorded* output of black tin exceeding 1,000 tons and none of these has been worked on a scale large enough to make them of first-class importance. Nevertheless the district has in the aggregate produced a considerable tonnage of tin. The Geological Survey estimated that the output of black tin during the period 1852 to 1905 exceeded 17,000 tons, since when further production has probably raised it to at least the 20,000-ton mark. If to this figure is added the obviously large though unrecorded output prior to 1852 it will be realized that the area's total production must amount to a very considerable figure.

Although the writer does not take an optimistic view of the future of the Wendron District as a whole, it does seem that there are still good prospects in at least one portion of it and, with a view to emphasizing these and as a means of giving a general picture of the area, it is proposed to deal briefly with the more important and interesting properties that it contains.

Section 3: The Wendron District

Wheal Trannack

Commencing at the extreme south-western corner of the area the first mine of note, and the only copper producer of any magnitude, is Wheal Trannack. The property is situated on the junction of killas and granite and at the time Henwood was writing in 1843 a narrow vein was being worked for copper in both rocks, a depth then having been reached of 84 fathoms. Tin was present in the upper parts of the lode, but downwards in granite the only yield was copper.

Collins states that the production of copper ore from 1822 to 1833 was 8,194 tons and that £20,000 profit was made before 1843. In this early working the mine was drained by means of a water wheel, but the power available in summer was insufficient to keep the mine drained to the bottom. In a later re-working in the '60's on a North lode only 115 tons of copper ore are recorded as having been sold.

Trumpet Consols

Half a mile north-east of Wheal Trannack and about half a mile south of Wendron church on the Helston-Redruth road are to be seen the remains of the Trumpet Consols mine. The property was formed by the amalgamation of the following small setts:- Old Trumpet, Wheal Dream, Wheal Noon, Wheal Valls, and Wheal Widden. These mines, which are entirely within the granite, had by 1870 attained a depth of 210 fathoms (adit 10 fathoms below surface) and they are thus the deepest workings within the district. The lodes are small, but were unusually rich. Henwood records numerous details of one which varied in width from one inch to one and a half feet. This lode, striking 22° S. of W., is the westerly prolongation of that worked in Wheal Ann (later Trumpet United), an old mine further to the east that is said to have reached a depth of 170 fathoms and which was very productive in its early days.

The only recorded production of Trumpet Consols covers the years 1854 to 1880, when 4,292 tons of black tin were sold. In view of the fact that these mines are very old and were being extensively worked

Section 3: *The Wendron District*

long before 1854 it is possible that their total output is as large or even greater than that of any other mine ever worked in the Wendron District.

About 25 years ago a proposal was made to re-open these properties, but in view of their long history it is highly probable that they would have been found to be completely exhausted and it is perhaps fortunate that the would-be promoters failed in their aim.

Trevenen and Tremenheere

A little to the south-east of Trumpet Consols and lying to the west of the Helston — Falmouth road is the very old Trevenen mine, re-opened in about 1860 and worked in conjunction with Tremenheere. Early in the last century Trevenen had reached a depth of 1,000 ft. and was then considered to be a deep mine. It seems to have been very rich in tin and profitable in its earlier years. Carne has recorded that the lodes were richest at a depth of 150 fathoms from surface. In the later working of these properties only 730 tons of tin are recorded as having been sold during the years 1854 to 1876. Tremenheere ultimately reached a depth of 204 fathoms.

The Lovell Group

To the north-east of Trevenen are numerous small mines scattered over an area of rather more than one mile square. The name Lovell is incorporated in the title of many of these small setts, but, with the exception of Old Wheal Lovell, none of them has achieved any considerable depth or output. Old Wheal Lovell — known also as "The Lovell" — is situated on either side of the main Helston-Falmouth road just north of the hamlet of Manhay. As long ago as 1864 it had reached a depth of 174 fathoms (144 below adit). During the period 1852 to 1896 production amounted to 1,526 tons of black tin, to which must be added the output prior to 1852 which is thought to have been very considerable.

Section 3: The Wendron District

Dividends totalling £13,330 were paid during the years 1846 to 1856. In its latter years the mine was worked in conjunction with the Mengearn and Combellack setts further eastward. Cape's Lode in Wheal Lovell, which was productive to a depth of about 150 fathoms, is said to be the same lode as that worked much further west in Wheal Ann and Trumpet Consols. According to Foster the North Lode is a wide one, 10 to 15 ft., and is very "bunchy".

East Wheal Lovell. — On the high ground just beyond the top of the steep hillside on which Old Wheal Lovell is situated, and three-quarters of a mile farther north, is the site of the celebrated East Wheal Lovell, perhaps the most famed of all the Wendron mines. The property consists of two setts — Tregonebris on the west side of the main road and Fatwork on the east. Collins states that in the early days two very narrow lodes, which appear to junction westward, were worked by the old men to a depth of 17 fathoms. The mine was re-opened in 1857, but at first operations were unsuccessful; later some very rich "carbona" ground was discovered in the Fatwork part of the sett and the venture became exceedingly profitable, over £38,000 being paid in dividends on "calls" of less than £7,000.

Foster, in his detailed description of the mine, states: "The lodes are usually narrow, sometimes a mere joint or line of division in the rocks, but occasionally a couple of inches thick; they consist of quartz, a litle clay, and red oxide of iron, and *per se* are utterly valueless. In some places, however, you get curious deposits of tin on both sides of the vein, occupying very little space but extremely rich, consisting of kaolinized granite containing much gilbertite and schorl". Foster added: "The East Wheal Lovell pipes and bunches resemble in their mode of occurrence some of the carbonas of St Ives, but differ in the absence of tourmaline which is very abundant at St Ives."

From very small workings tin to the value of £39,000 was sold in a period of 20 months from October, 1869, to May, 1871, yielding a profit of £27,000. At one time a drive in one of these ore-bodies was valued at £1,000 per fathom. The main shoot of ore was followed from the 40-fathom level down to the 110 as one continuous pipe and was in the shape of a long irregular cylindroid with an elliptic base, generally about 14 ft. long by 7 ft. wide. There were several other pipes and bunches of similar character. The mine was ultimately sunk to the 127-fathom level and finally abandoned in 1901, although after 1891 production was

Section 3: The Wendron District

very small. The Geological Survey gives the output during the period 1859 to 1891 as 2,405 tons of black tin.

In 1928 a half-hearted attempt was made by the London Tin Syndicate to re-open the mine. In the western part of the sett the old Colonel Shaft was cleared and unwatered to 40 fathoms from surface and a little development and exploration carried out. All motive power was obtained from a single diesel engine, but, following a breakdown of this machine, all work was abandoned. A little shallow prospecting was also performed considerably further eastward and at the time there was some talk about sinking a new "central vertical shaft" to open up all the ore-bodies in the locality. One might as well look for the proverbial needle in a haystack as expect to discover more of these small and erratic ore-bodies by sinking a central vertical shaft!

Wendron Consols

Immediately north and east of Wendron Village and lying along the banks of the River Cober is the Wendron Consols mine, which was formed by an amalgamation of the Trenear and Ball Dees setts. There are several lodes in the property and these have two main bearings — one series striking nearly E. and W. and the others having the typical W.S.W. bearing of the district. Operations here commenced in about 1852 and by 1864 six lodes were being worked, a depth of 72 fathoms having been reached from surface. (There is no adit). Up to this date dividends of £8,960 had been paid on an expenditure of £4,645. The recorded black tin sales for the period 1854 to 1869 amount to 2,390 tons, in addition to which 48 tons of copper ore were sold.

Basset and Grylls, or the Porkellis Mine

Underneath the Porkellis Moor and extending northward to within a short distance of Porkellis Village is the Porkellis mine, which has been

Section 3: The Wendron District

worked at times as "Basset and Grylls." This is, in many respects, the most important mine in the whole area and its production of tin may well have exceeded that of any other mine in the Wendron hills. Porkellis has had a long, though chequered, career, having been many times abandoned and re-opened. It was last working in 1938, but was then literally thrown away in a panic by the sole owner at a time of slump in tin prices. Reference was made to the property in the first part of Section 2 (see above pp.23 - 24) and it is very much to be regretted that the authorities did not re-open this relatively small and shallow mine when the national need for tin was so acute. If a fraction of the money (and labour) which has been so uselessly expended in Cornwall and Devon had been concentrated on re-equipping and further developing a single small mine such as Porkellis it is certain that far more useful results would have been obtained.

In the early days the mine was worked beneath the alluvials of Porkellis Moor, but in about 1859 the decomposed granite above the shallow workings caved in and the mine was filled with a rush of mud and water. Seven men lost their lives in this disaster and as this part of the mine was abandoned the bodies were never recovered. The only returns of black tin available up to this date are those covering the years 1852 to 1860, when 1,278 tons were sold. Following the loss of the old mine the property was united with several adjoining setts as Porkellis United, and later as Basset and Grylls and operations were commenced afresh on several lodes about a third of a mile further north. This working lasted until about 1880, the total recorded sales of black tin from 1852 to 1880 being 4,323 tons.

During the early years of the present century the dumps left from the last company's operations were milled and the recovery of tin was sufficiently high to attract renewed attention to the mine. It was found that the particular dumps in question, although deposited near Tyack's Shaft (the site of the last company's principal operations), had actually come from some old workings further up the hill and closer to Porkellis Village. A company was subsequently formed in about 1907 to re-open these workings and it is this part of the sett with which we are now principally concerned.

There are two main lodes in this part of the mine — the "Old Men's" and "Wheal Cock." Operations at first commenced on a crooked inclined shaft sunk on the latter lode, but were then abandoned in favour

Section 3: The Wendron District

of the "Old Men's" perpendicular shaft, which passes through the north-dipping lode of the same name at about 35 fathoms below adit. For some years work was mostly confined to the Old Men's Lode, which was developed to a depth of 75 fathoms below adit (adit 75 ft. from surface) and all operations were on a small scale until a new mill, nominally of a 100-ton per day capacity, was erected in 1919. The great slump in 1921, however, brought the enterprise to a close and, although the plant was maintained intact, nothing further was done until the mine was unwatered for a brief period in 1926 in order that it might be sampled by interested parties. The water was then again allowed to rise, but in 1927 the property was re-opened under the title of Jantar (Cornwall) Ltd.

Operations were then conducted in a much more vigorous manner than hitherto. Electric power was brought in and utilized for all purposes and the mill augmented and modified. The increased pumping power then available made it possible to resume the development of the Wheal Cock Lode in depth, which had hitherto been impossible because of the small size of the gas-engine-driven Cornish pump. The water collected at and pumped from the 75-fathom level varied from over 300 gallons in winter to 150 gallons a minute in summer, the inflow being approximately evenly divided between the two lodes. Incidentally, when at a later date the mine was sunk to the 90-fathom level, the water at that depth proved to be so little that it could be handled by a 3-in air pump. The mine responded excellently to the expenditure that had been made underground and at surface, but the Company was destined to a short life, the world-wide slump in tin in 1930 bringing all operations to an end in October of that year.

In March, 1934, a private company, Porkellis Tin Mines, Ltd., was formed to take over the property, which was immediately unwatered and work actively resumed. The Wheal Cock Lode in the olden days had been worked by the old men down to the 35-fathom level, but, for some reason, it had then been abandoned. The development of this lode in the later workings extended down to the 75-fathom level with highly-encouraging results, values ranging from 30 lb. to 70 lb. per ton of black tin being not at all uncommon in some parts of the lode and even 200 lb. and over were encountered at times. In addition an exceedingly rich carbona-like deposit formed by a number of "droppers" from the foot-wall of the Wheal Cock Lode was discovered between the 45 and

Section 3: The Wendron District

60-fathom levels and this produced a large amount of tin within a short period. The average lode width was about 4 to 5 ft., but in places it widened out greatly. The maximum length of the developments on this lode were about 1,200 ft. in the middle section of which a westerly-dipping ore-shoot approximately 750 ft. long was almost entirely stoped out from surface down to the 75-fathom level. The Old Men's Lode, which is situated further south, is on the whole a lower-grade ore-body averaging about 1% of black tin over a width of 4 to 5 ft.; it is, however, more regular both in width and in value than the Wheal Cock Lode, which is "bunchy" in both respects. The Old Men's Lode, which is a red one and of a very soft nature, has likewise been developed laterally for a maximum distance of about 1,200 ft. and an ore-shoot approximately 850 ft. long stoped out down to the 75-fathom level. Both lodes have been worked further westward by the old men; indeed the last company accidentally holed to some shallow western workings on the Wheal Cock Lode, in consequence of which the new mine was nearly flooded. In view of these old workings any future exploration in that direction should be conducted with great caution. Eastward the Old Men's Lode has apparently been worked in a shallow open-cast working 1,000 ft. east of the Old Men's Shaft. It would seem that both lodes extend for a very considerable distance along their strike and that the workable values are concentrated in a number of ore-shoots, one of which on each lode constituted the stoping ground in the late Porkellis workings and other shoots further westward gave rise to part of the old men's workings under the moor. If the most recent developments were extended further eastward there would seem to be an excellent chance of picking up still further shoots in that direction.

As a consequence of the rich discoveries made during this last period of working considerable profits were realized and the company was very prosperous for a while. Unfortunately, however, from the time of the re-opening in 1907, the mine had been operated in a very short-sighted manner, every rich bunch being hastily exhausted without any concerted attempt being made to develop the property regularly and systematically. Disagreement later ensued between the three individuals comprising the company on matters of development policy, and, ultimately, one of the principals bought out the interest of the other two partners and continued the working of the mine on his own account. Ill-fortune, however, again intervened, for the individual who was now

Section 3: The Wendron District

the sole owner was embarrassed by a short-lived slump in metals that occurred soon afterwards and he was unable to supply additional working capital for pressing developments that should have been carried out before. During this period of depression the mine had to live on its own resources and the manager succeeded in sinking and equipping a new "lift" entirely out of revenue. A sump winze was sunk on the Wheal Cock Lode from the 75 to the 90-fathom levels and a cross-cut extended south to the Old Men's Lode and to the shaft. Two rises were put up — one on the Old Men's Lode and the other to the shaft sump, the latter rise being timbered and brought into commission as the new part of the shaft in November, 1937.

At the 90-fathom level the Wheal Cock Lode was poor eastward and, as there was no money for its further development, work was concentrated on the west drive. There, after passing through the crosscourse which is only a few feet west of the shaft and which faults the lode 15 ft. to the south, the ore-body was found to be 6 ft. wide and worth about 30 lb. per ton and as such was driven on for 150 ft. and stoping commenced. On the Old Men's Lode at the 90-fathom level an ore-shoot about 600 ft. long was developed in ground worth about 1% over a width of $3\frac{1}{2}$ ft. The then manager told the writer that he estimated that 20,000 tons of 1% ore had been blocked out on the Old Men's Lode below the 75-fathom level, although some of this was stoped and a certain amount of broken ore still remains in the stopes. In addition to these developments a small hard blue lode that nevertheless carried high tin values was encountered in the 90-fathom shaft station and this was again seen in the shaft rise and in a cross-cut extended south to intersect it at the 60-fathom level. Above the 90 this lode was split into branches and was of irregular value, but its further development laterally and in depth is thought to be worth while.

Notwithstanding these encouraging results at the bottom of the mine the lack of adequate development over a number of years had made it impossible to carry on during the slump without incurring losses and this the then owner was unable to face and he decided to abandon the property forthwith. Before operations ceased a well-known mining engineer inspected the mine and it is known that he took a favourable view of the prospects. The Old Men's Lode dips north at about 60° and the Wheal Cock Lode, although erratic in dip, is, on the whole, dipping steeply southwards. The engineer who inspected the property was of the

Section 3: The Wendron District

opinion that the two lodes would join about 90 ft. below the 90-fathom level and he thought that the prospects of enrichment at the point of junction were sufficient to warrant the immediate sinking of a couple of winzes to prove the property 90 to 120 ft. below the bottom level. The work was estimated to cost £2,000 and the engineer in question was strongly of the opinion that the property merited that additional development in depth. However, the owner was unable to finance the business any further and he decided on immediate abandonment. It can only be said that the precipitate dismantling of the Porkellis Mine is one of the most extraordinary and regrettable incidents in the history of Cornish mining.

As far as the future is concerned it is obvious that if ever the mine is re-opened it will have to be sunk immediately to the depth necessary for at least two new levels, as everything of value has been stoped out above the 75-fathom level; the 90-fathom level alone could not long sustain production at anything like the previous rate of milling which varied from 80 to 120 tons per day. It has to be admitted that in the Wheal Cock Lode especially the cassiterite was becoming very much finer in depth and this fact is suggestive of the lower zone of that lode having been reached. It must be admitted too, that past experience in this district does not encourage the hope that any of its lodes will continue productive to any great depth. Nevertheless, the results so far achieved at Porkellis do suggest that a further 200 ft. of sinking would be a worth-while speculation. Quite apart from the possibilities in depth and the very good prospects of discovering new ore-shoots by additional lateral development along the existing lodes there are the chances of opening up other parallel ore-bodies. Mention has already been made of one at the 90-fathom level and it is known that a further lode was cut by diamond-drilling northward some years ago although it was never developed. Several other lodes exist south of the Old Man's Lode; these were mined to a limited extent down to a maximum depth of 75 fathoms by the old company, which commenced work in 1859. Although these southern workings were previously thought to have been exhausted, the surprising discoveries made since 1907 by developing the Old Men's and Wheal Cock Lodes deeper than that attempted by the old workers makes it an interesting speculation whether these other workings, too, would also repay the cost of unwatering. It is noticeable on the plans that the old company does not seem to have done any driving on these southern

Section 3: The Wendron District

lodes west of the cross-course, which lies just west of the Old Men's Shaft. In view of the good values found in both the Wheal Cock and Old Men's Lodes after the cross-course had been penetrated it seems an additional reason for exploring these southern lodes further westward. The late company's pumps, however, were of insufficient power to risk cutting any additonal water and so very little work was done south of their shaft.

Porkellis is a shallow mine, the actual depth of the bottom workings, nominally 90 fathoms below adit, being only about 560 ft. from surface. The mine is equipped with a perpendicular shaft and electric power is available close to that shaft. In view of these facts the unwatering of the mine should not be a costly matter and the development necessary to prove whether it merits complete re-equipment and re-working should not entail heavy expenditure. The mine is not a high grade one; it has only averaged about 1% of black tin in the past, but the ore is cheap to mine and simple to mill and consequently overall costs can be kept relatively low. In the past they averaged about 21s per ton. Of all the mines in the Wendron area Porkellis seems to be the one most worthy of further development and it is strongly to be hoped that this work will be undertaken at some future date.

Since the abandonment of the mine in 1938 a local landowner has purchased the mineral rights with the avowed intention of preventing the mine from being re-worked. When in operation the red muddy water discharged from the mine and mill coloured the waters of the River Cober, running through his estate, and he wished to prevent this recurring in the future. However, the Mines (Working Facilities) Act of 1934 made provision for just such cases of obstruction on the part of a mineral owner and there is little doubt but that such a prohibition could be nullified and anyone wishing to re-work the mine could obtain power to do so.

Garlidna

Approximately a quarter of a mile east of the Porkellis mine and lying beneath the eastern arm of the Porkellis Moor is the old Garlidna

Section 3: The Wendron District

property, which was worked somewhat extensively about 1840 or earlier with Wheal Ruby, and, at a later date, with the Polengrean Sett, under the title of Garlidna United. Garlidna is known to have been sunk to a depth of 70 fathoms, but being situated beneath the water-logged moors the water in the mine was heavy. These setts contain numerous lodes, but although several times re-worked, they do not seem ever to have been very successful. The only recorded sales of mineral show 136 tons of copper ore in 1847 and 174 tons of black tin and 3 tons of copper ore between 1861 and 1880. Some small-scale operations were again carried on at Garlidna during the present century, but the mine was finally abandoned in 1917.

Wheal Enys

A few hundred yards north-west of Porkellis Village can still be seen the old pumping house of Wheal Enys, which Collins states was known at an earlier period as Wheal Vernon. According to the same authority this very old mine, which was worked before 1815, was sunk to a depth of 60 fathoms and at one time gave promise of much extension. A proposal for re-opening in 1907 did not mature. The published sales of black tin are 259 tons between 1853 and 1859. An elvan dyke runs through the sett.

Boswin and Wendron United

At the top of the hill, about a quarter of a mile north of Porkellis Village, there is a group of small mines that has yielded a certain amount of tin, although it is doubtful whether any of them has ever been a commercial success. The last of these mines to be worked was Boswin, also known as Wheal Puffet. This property, which is an old one, was re-started by Messrs. H. S. and E. Gordon about 1909 and worked to a depth of 60 fathoms. Two principal lodes were worked and several lesser intersected.

Section 3: The Wendron District

Apparently the old men had left intact a certain amount of payable ground and a small mill was erected and a modest tonnage crushed, but the venture did not last long and it was finally abandoned in the early part of 1915.

Immediately east of the Boswin Sett is the Wendron United mine, to which the workings east of the shaft on the Boswin Main Lode holed. Wendron United was sunk to a depth of at least 86 fathoms and was worked upon six lodes of which the Geological Memoir of the district gives several particulars. The only recorded sales of tin are a few tons between 1853 and 1876.

Balmynheer

Half a mile north-east of the Wendron United Mines is Balmynheer, an old mine that is said to have been worked in its early days to 70 fathoms below adit (adit 10 fathoms from surface). The chief feature of interest is the "Hownan" Lode, a remarkable carbona-like mass that was described by Foster in the following terms:-

The deposit consists of a large irregular mass of stanniferous altered granite. A vein or slide about 6 inches in thickness, consisting of white clay and a little quartz and mica, underlies N. 60° and bears E. 32° W. Below the slide the irregular mass of tin-bearing rock varies in thickness from 30 to 50 feet, and underlies in the same direction as the slide. Between this mass and the granite there is no regular plane of separation. Occasionally there is some tin ore above the slide. This mass, which consists of quartz, chlorite, gilbertite, iron pyrites, zincblende, and tin ore, and occasionally a little wolfram, extends for 36 fathoms along the strike of the slide, the lowest workings being at 30 fathoms from surface.

In 1873 it was reported that the lode at the 20-fathom level was productive for 40 fathoms in length and for 10 fathoms in width, yielding an average of 28lb of black tin per ton. In 1876, 2,200 tons of tin ore was stamped, yielding over 1% of black tin per ton. During the last working, which commenced in 1864, operations were mostly confined to the Hownan Lode and the mine was only unwatered to the 30-fathom level. The water in winter-time was heavy and with frequent trouble

Section 3: The Wendron District

from decomposed granite in the adit the mine was sometimes flooded. It has always been worked on a small scale without adequate capital and it is questionable whether the Hownan Lode has ever been adequately explored.

Calvadnack

A quarter of a mile north-west of Balmynheer is the Calvadnack mine, which was started in 1850 and which was worked somewhat extensively for a time on a series of lodes parallel to those of Balmynheer, the most important of these being the Grand Jack Lode. By 1864 the mine had reached a depth of 80 fathoms. The recorded production shows an output of 1,150 tons of black tin during the period 1854 to 1875, but the working of the mine resulted in a considerable loss. Until the commencement of the neighbouring Polhigey mine it was the tradition in the district that Calvadnack was a rich property. The bottom was said to be like the proverbial "jeweller's shop," but there were few signs of any such riches when the workings were unwatered to a depth of 300 ft. by Polhigey! An old friend of the writer, who knew the Wendron mines well, used to say that the Calvadnack lodes were hard blue "hungry" things and that in order to get the little bunches of tin that they contained it was necessary to stope the whole of them and then they were no good at all!

Polhigey

Immediately to the east of Calvadnack and worked on the same series of lodes is the Polhigey mine. A little work was done here in 1862 to 1864, but when prospecting commenced in the early '20's of the present century the sett was practically virgin and the only workings to be seen were long trenches made by the old men along the outcrop of the lodes. The most recent operations commenced when two Redruth mining engineers took up the sett and sank a small prospecting shaft at the eastern end of the

Section 3: The Wendron District

property. The sinking was only continued to a depth of 75 ft., but from that horizon a few short lateral developments disclosed excellent values. The individuals who were financing the work felt unable to carry it any further entirely on their own account, but they were ready to join in with any larger interests who were prepared to open up the property on an adequate scale.

In 1926 the London Tin Syndicate decided to take a hand in the matter and to develop the property extensively. A company, entitled Polhigey Tin Ltd., was accordingly registered with a capital of £160,000, and it was decided to enlarge the existing prospecting shaft and then to sink it to a depth of 300 ft. New levels were established at 175 ft. and 300 ft. from surface and an adit driven that gave free drainage down to the 75-ft. level. The property was found to contain numerous lodes; it was reported at the time that there were 13 dipping south and one (the principal lode) dipping north. The latter is the eastern continuation of that worked in the Calvadnack mine as the "Grand Jack" Lode and it was on this that the bulk of the mining was subsequently done in Polhigey. Some of the other lodes were also opened up, but the values in them were patchy. At the commencement of the development programme some excellent values were disclosed, but before there had been sufficient work done to assess the value of the sett thoroughly the decision was made to sink a large 4-compartment shaft further westward and to erect a 200-ton-a-day mill and diesel-driven power plant.

It had been intended to sink the new main shaft, named Roberts', to a depth of 600 ft., but the greatest depth reached was 420 ft., at which horizon the fourth and deepest level was established. Considerable cross-cutting was done and the lodes were extensively explored along their strike and connexions made with the old Calvadnack mine to the west at the 175 and 300-ft levels. It can only be said, however, that the mine was a disappointment and it completely belied the expectations which were founded on the early developments. There were admittedly some good "bunches" in the lodes, but of the tonnage mined it is probably correct to state that the average recoverable value was not more than $\frac{1}{2}\%$ of black tin per ton of ore. The ore may have contained nearly double that amount of tin and, of course, it has to be admitted that the mill was not an efficient one, but it proved impossible to recover more than $\frac{1}{2}\%$ of tin. These results remind the writer of his old friend's previously-quoted remarks about Calvadnack and, as that wise old man once added: "As

Section 3: The Wendron District

Polhigey is only the eastern extension of Calvadnack what else can one expect!"

Whatever may be thought of the upper levels in Polhigey it is certain that at the bottom of the mine the lodes were even poorer and were of a very "hungry" appearance. In the writer's opinion there is every indication that the ore-bodies were of a very shallow nature and, as in so many other mines in the area, were failing in depth. Although, as previously stated, the mine gave early promise of much better results, a mistake was undoubtedly made in spending so much capital on surface equipment before it had been possible to arrive at a true estimation of the value of the property. Milling commenced in the early months of 1929 and although costs were certainly low, it is doubtful whether receipts ever equalled expenses. With the advent of the world-wide slump in tin in 1930 it proved impossible to carry on any longer and milling ceased at the end of October in that year. The published sales of black tin during 1929 and 1930 amounted to 286 and 256 tons respectively. The plant was maintained intact for a while, but later, in common with all the other mines of the Anglo-Oriental group in Cornwall, Polhigey was abandoned and the plant dismantled. Thus ended the most ambitious attempt in recent years to work a Wendron mine on a comparatively large scale.

Conclusion

This review of mining in the Wendron area is now concluded with the exception of a few small and unimportant setts which do not merit attention. Although the mines in the area are all comparitively shallow and although there are many miles of country where there has been no exploration in depth, there do not seem to be many prospects that are worthy of further trial. Collins, amongst other authorities, thought that there were possibilities of great expansion in this district, but the writer does not see that there are any facts to support such a view. The Porkellis mine, and that alone, seems to offer a reasonable prospect of a return on any further expenditure of capital if worked on a moderate scale — say, 100 tons per day. It is sincerely to be hoped that someone will tackle

Section 3: The Wendron District

this property again in the not distant future. Of the other mines in the area little or nothing can be expected.

Section 3: The Wendron District

Section 3: The Wendron District

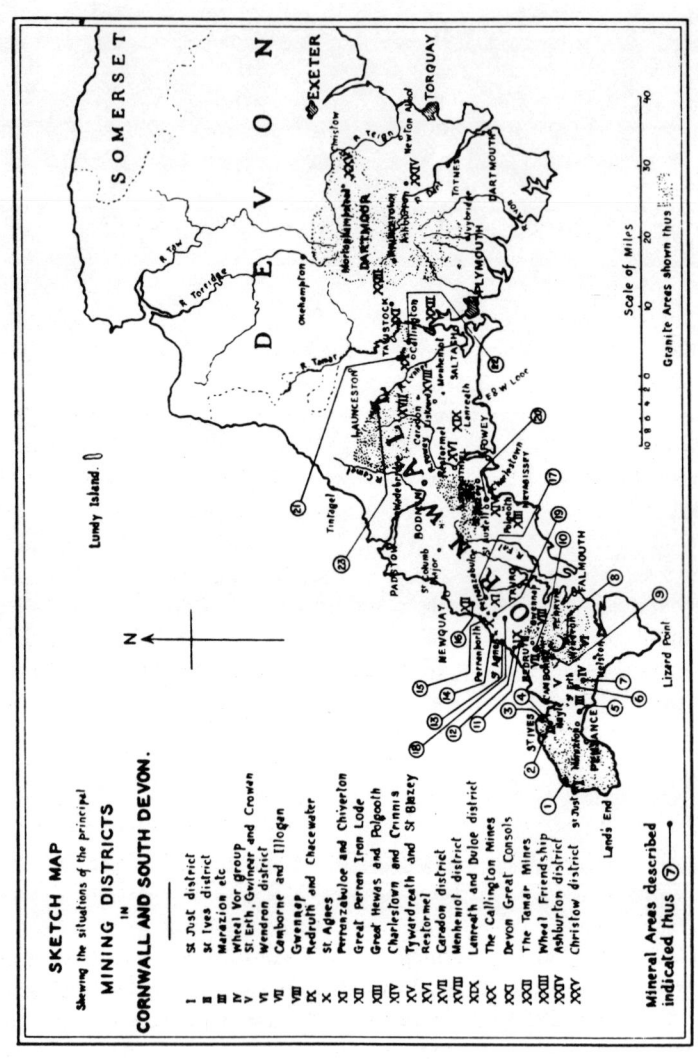

SECTION 4: SOME USEFUL PROSPECTS FOR THE FUTURE

Section 4: Part I

LEVANT TO SOUTH CONDURROW

In a series of articles which appeared in the *Magazine* during the recent war the author dealt with some of the Western mining districts of Cornwall. The growing pressure on base-metal supplies, however, is such that it has been thought opportune to prepare a review of some of the mineral possibilities latent in the county. The notes which follow, prepared originally for private circulation among members of the Cornish Mining Development Association, of which the author is chairman, are thought worthy of wider study. They are the outcome of a great deal of investigation and research extending over nearly 30 years. No claim is made that they cover all the favourable prospects but merely that they indicate some of the areas most worthy of investigation. Although no reference has been made to the Caradon mining district, north of Liskeard, the south-eastern portion of that area merits attention if ever the demand for copper again makes the mining of the red metal attractive in Britain.

(1) Levant Mine, Pendeen

This famous old mine was once one of the most important tin and copper producers in the county, from its commencement in 1820 until the abandonment in 1930 the recorded output amounted to approximately 25,000 tons of black tin, 136,000 tons of copper ore, and a considerable quantity of arsenic. The mine has been worked at and about the junction of the granite and mixed killas-greenstone rocks, but the ore occurs principally in the last-named. The workings extend under the sea for a

Section 4: *Levant to South Condurrow*

length of $1\frac{1}{2}$ miles and from surface to a maximum depth of about 2,200 ft.

The ore-shoots at Levant dip north-westward under the sea, but at a much lesser angle than the killas-granite contact. There are numerous lodes, some of which have only been partly developed and under the sea there is still considerable scope for lateral exploration. Some of the lodes are continuous with those of the neighbouring and active Geevor mine — eg. the "Prince of Wales" lode of Levant is the "No. 3 Branch" lode of Geevor and Levant "North" lode is the "No. 2 Branch" lode in Geevor. In Levant the latter ore-body junctions under the sea with the "South" lode at an acute angle, thus forming the principal ore-shoot of the mine, 3,000 ft. or more in length.

In its latter years the Levant mine was principally a tin producer, but even in the deepest workings — such as the sump winze west of the New Submarine shaft — copper was still in evidence in places. Indeed there does not yet appear to be any indication of the lower limits of the tin zone having been approached in the deep under-sea workings. Over a considerable period of years the average recovery of the old and inefficient mill was at least 37lb. of black tin per ton — a figure much in excess of that obtainable at most other Cornish mines in recent years, while during the last two years that the deep part of the mine was in operation the recovery varied between 42.4lb. and 52.6lb of black tin per ton.

During the 19th Century the conduct of operations at Levant was a classical example of improvidence on the part of a "cost book" or unlimited liability company that had reaped enormous dividends from a trifling initial outlay without making proper provision for the future. Consequently the deep under-sea part of the mine could only be reached by means of subsidiary submarine shafts and the ore had to be rehandled so many times that working costs were quite unnecessarily high. The proposal to sink a new shaft giving direct access and ventilation to the deep under-sea workings was repeatedly shelved on the score of initial cost. Owing to the lack of such a shaft the miners still had to be raised and lowered by the "man-engine", the breakage of which in 1919 caused a very heavy loss of life and made it impossible to continue the working of the deep part of the mine.

The company was subsequently reconstructed and, until such time as the money could be found for the sinking of a new shaft, the mine

Section 4: Levant to South Condurrow

was abandoned below the 190-fathom level (approximately 1,290 ft. from surface) and the available resources concentrated on the further exploration of the upper levels. These so-called worked-out parts of the mine responded to such good effect that during the ensuing ten years some 3,000 tons of black tin were produced. The milling recovery during this period (when much dump material was also crushed) averaged about 27lb. of black tin per ton. However, the reconstructed company, short of capital, was unable to sink the new shaft, the miners having to climb to and from their work daily — a vertical distance of about 1,200 ft. Under such conditions it was impossible for the concern to build up any reserves and the great slump in 1930 found it unable to carry on. The company had earlier obtained a loan from the Government through the Trade Facilities Committee, but as it was unable to meet its obligations the Treasury foreclosed and appointed a Receiver.

Owing to the impervious nature of the killas-greenstone rocks the Levant mine was singularly dry, the inflow of water being only about 60 gallons per minute. Somewhat strangely, too, the mine had a particularly good health record; indeed a well-known medical man told the writer that he had never known a man to have silicosis who had worked solely in Levant. The air temperature, however, in some of the lower workings was excessive. If the mine is ever worked again, at still greater depths, the best solution of the ventilation and ore-handling problems may well be the sinking of a major inclined shaft to give direct access from surface to the deep under-sea parts of the mine.

The Pendeen–St Just area possesses other possibilities than those embraced by the Geevor and Levant mines. This is especially true of the further seaward development of the adjacent Botallack mine lodes and also those of the small Wheal Bellan and Wheal Hermon properties, half a mile south of Cape Cornwall. Much capital has been wasted on these mines (especially at Botallack) in attempting to rework the landward portions of these lodes, but seawards the prospects in some of them are still good.

Section 4: Levant to South Condurrow

(2) Wheal Racer, St Ives

This is a small mine on the western slopes of the very prominent granite mass of Rosewall Hill, the situation of the surface workings being approximately Lat. 50° 12' and Long. 5° 31' W. As so far developed the mine contains two lodes, these appearing to be on the western extension of the series once so profitably worked for tin in the Rosewall Hill and St Ives Consols mines further east.

The two lodes in the mine are reported to carry fair values and to be capable of being very easily and cheaply mined. The existing workings are about 360 ft. deep and it is thought that the two lodes will junction about 60 ft. below the present bottom of the mine and such junctions frequently produce important ore-bodies. As there is no adit the mine is flooded to surface, but the amount of water to be pumped is insignificant and was insufficient to supply a small mill when the mine was in operation about 30 years ago. Informed local opinion considers that though small this is an excellent prospect.

(3) Wheal Trenwith, St Ives

Extending south-west from the town of St Ives there is a series of lodes which have been considerable producers of tin, copper, and uranium ores. The principal mines worked on these lodes are, from east to west Trenwith, St Ives Consols, and Rosewall Hill and Ransom United. With the exception of Trenwith, which is on the contact of the mixed killas-greenstone rocks and the granite (as at Levant), these mines have all been worked in granite. The tin production, amounting to over 17,000 tons, came almost exclusively from the two western groups in the granite, whereas the copper and uranium were confined almost entirely to Trenwith in the killas.

In 1909 work on the whole of this group of properties was resumed by the St Ives Consolidated Mines, Ltd., and the greater part of the mines unwatered. Apart from Trenwith, the eastern section, there would appear to have been little justification for this scheme for it was evident from past records that these mines were virtually worked out. At Trenwith, however, there was every prospect of a successful tin mine being developed by deeper sinking in granite under the copper zone. These possibilities were confirmed when the mine was unwatered and some very good tin values were discovered in the lower levels, the deepest of which is only 600 ft. below adit.

In the past the mine had been worked entirely for copper, but the value of the ore was seriously reduced by the large amount of pitchblende present. By 1909, however, this "impurity" had become a very valuable asset and a treatment plant was erected and a considerable tonnage of ore left by the old workers mined for its uranium and radium content. This ore came mostly from the "Standard" lode, which is the principal one of these mines, but there are others yielding uranium which have been only partly explored. The author understands, however, that the percentage of pitchblende was decreasing in depth and the company resolved to develop the mine in future as a tin proposition. Unfortunately the 1914-18 war had by then broken out and this resulted in an immediate loss of 40% of the labour force, many of whom were naval reservists. As the company were also operating the nearby Giew mine, from which the bulk of their production was being obtained, it was decided, as a temporary measure, to close down Trenwith and their other properties and to concentrate all available labour at Giew. The amount of water to be handled at Trenwith was very little and it was felt that unwatering there in future would present no difficulties. Unfortunately the company foundered in 1922 during the acute slump in tin prices then prevailing and Trenwith, in common with their other properties, was dismantled and abandoned.

Beyond treating a lot of uranium-bearing ore and confirming the possibilities of tin being found in depth the last working did nothing to prove the property and its vigorous prosecution for tin is still well justified.

(4) Giew Mine, Towednack

This mine forms a part of an extensive group of small mines known as Reeth Consols which have been worked on the southern and western

Section 4: Levant to South Condurrow

flanks of the lofty granite Trink Hill. The records of production are very incomplete, but it is known that the sales of tin up to 1867 had realized £234,000. The portion known as the Giew mine was, as already noted, re-opened by the St Ives Consolidated Mines, Ltd., about 1909 and sunk from a previous depth of 850 ft. to 1,300 ft. below adit level. There are two lodes in the mine, but only one was developed. This was extensively worked for 1,200 ft. or more east of the main shaft to a point approximately under the summit of the hill. Over a number of years the average grade of ore was 26 lb. to 27 lb. of black tin per ton, but much dilution occurred in an ill-advised attempt to keep the mill crushing to full capacity.

From the commencement the new company was seriously handicapped by a shaft that was very crooked as well as being inclined, but before the mine could be soundly established the 1914-18 war broke out and thenceforth operations were crippled by shortage of labour and the rise in the cost of supplies. A new shaft on the eastern side of Trink Hill was urgently required, especially from the point of view of ventilation, which was very bad. During the slump of the early 20's the concern continued to struggle against overwhelming odds at a time when practically every other Cornish mine had had to suspend production. Towards the latter part of 1922, however, the company collapsed and the mine was abandoned. About 1928 a proposal was under consideration for re-opening these mines and sinking the necessary new shaft further eastward, but the renewed slump in tin in the 30's put an end to any further action in the matter.

The country to the east of these mines is entirely unexplored in depth; furthermore, from shallow trials and from accidental discoveries made during the sinking of wells, etc., it appears that this series of lodes exists for at least a mile east of anything yet done on them in Giew. In addition there are several other lodes in the area which are not even connected with the late Giew mine workings and which have long been considered worthy of further investigation.

Section 4: Levant to South Condurrow

(5) The Marazion Area

The Marazion mining district contains an extensive group of lodes which merge gradually into other districts. The present notes, however, are confined to a portion of the area about half a mile wide and extending from Marazion Marshes to the village of Goldsithney, a distance of approximately $2\frac{3}{4}$ miles. The rock formation here consists of killas, although certain elvan dykes are also present. The major granite mass of the Land's End outcrops $\frac{3}{4}$ mile to the north-west and the lesser granite dome of Tregonning Hill two miles to the east. In addition granite is also present one mile to the south at St Michael's Mount. Although the area does not come within the metamorphic aureole of any of the granite masses mentioned the presence of tin in the midst of what has hitherto been a predominantly copper district is suggestive that this may be the site of another emanative centre of some importance.

The principal mines in this district are, from west to east, Wheal Darlington, Wheal Crab, Wheal Virgin, West and East Rodney, Wheal Hampton, Tregurtha Downs, and Owen Vean. The first three of these were copper mines although some sales of tin were made. Their maximum depth was about 700 ft. below adit level. The remaining mines were all worked for tin, their total depth from surface varying from 260 ft. in the case of Wheal Hampton to a maximum of 720 ft. at Tregurtha Downs. The last-named property was worked on a modest scale during the closing years of the 19th Century, but the great volume of water to be handled (up to 1,500 gallons per minute) combined with the very low price of tin then prevailing made the venture unprofitable. Mismanagement is also said to have been an important factor. Of the three lodes worked the "South," which is 6 ft. to 8 ft. wide, is reported to have averaged about 22 lb. of black tin per ton, but the "North" lode, 3 ft. wide, and the great "Elvan" lode, which was sometimes very wide indeed, were both of considerably higher value.

Wheal Hampton, which is the youngest of these mines, was worked by a small company for a few years up until the closing months of 1914 when the difficulties resulting from the war added to a lack of capital brought the whole venture to a close. The only lode extensively worked was the "South" lode of Tregurtha Downs and that was extensively stoped from surface to the bottom level and for practically the

Section 4: Levant to South Condurrow

full length of the property — namely about 700 ft. The lode averaged about 30 lb. of black tin per ton and the values were well maintained to the bottom of the mine, the mill recovery during the 1910–14 period averaging 26.5lb. of black tin per ton of ore crushed. The average width of the lode appears to have been 3 ft. to 4 ft.; in some places it was very much wider. One of the principal reasons for the failure of the venture was that the workings were severely restricted in length by reason of the flooded Tregurtha and Rodney mines existing to the east and west respectively. Ultimately it was decided to undertake the costly task of unwatering these mines by erecting two large Cornish pumps, but the capital available was insufficient and operations ceased before any of the machinery could be completed.

A few years ago a proposal was made for the re-opening and working of all these mines on a large scale. To facilitate this it was intended to sink a new perpendicular shaft to the south (all the lodes dip in that direction) to an initial depth of 900 ft. Although the scheme did not materialise there can be little doubt that it was the correct policy as far as these mines are concerned. From the earliest days the working of this extensive area has been rendered unprofitable because the working of a single small mine entailed pumping practically all the water in the district. If these mines could be tackled as one major concern which would be capable of producing a tonnage of ore commensurate with the amount of water to be handled, the chances of success would be excellent. Furthermore the success of the initial developments in this area might ultimately lead to the area of operations being considerably extended so as to take in other mines on more or less parallel lodes.

(6) The Godolphin Hill Area

Two miles to the east of the Marazion district the granite again appears in a relatively restricted but economically very important outcrop, comprising the twin hills of Tregonning and Godolphin. Some of the richest tin mines ever discovered in Cornwall have been worked around the flanks of this intrusion and there still remains considerable scope for further discoveries.

Section 4: Levant to South Condurrow

The area may conveniently be divided into two districts. In the first place it is proposed to deal with the one that lies on the north side of Godolphin Hill, which forms the northern end of the intrusion. This district may be defined as an area extending from the village of Godolphin Cross on the east to Relubbus on the west — a distance of $2\frac{1}{2}$ miles. In width it extends rather more than a mile northwards from the summit of Godolphin Hill. The area contains numerous small tin mines which are deserving of further investigation, but the part of more immediate interest consists of the northern slope of the Godolphin Hill itself and the highly mineralized zone of killas at its foot.

The north side of the hill contains several lodes at and about the killas-granite junction which have produced a good deal of tin slightly further west in the West Godolphin mine. The systematic development of these lodes further eastward and immediately north of the centre of the hill has long been thought to be a first-rate speculation, but for some unknown reason it has never been tackled. The only work that has been done consisted of the driving of shallow adits and sinking of shallow pits, some of which gave encouraging results. The hill rises over 400 ft. above the valley at its northern foot and a really deep cross-cut adit could, with advantage, be extended for a mile through this most interesting piece of ground. This seems to be a prospect that should have been undertaken long ago.

Even more important, in the author's opinion, is the group of lodes under the valley at the foot of the hill. The most extensive workings here are those of the old Godolphin mine, which was an important copper producer as long ago as 1678. The mine was later worked by steam power and the workings extended to a depth of about 700 ft. before they were abandoned in 1847. When writing of the mines of this district in 1867 Captain Charles Thomas, a celebrated authority on Cornish mining and geology, stated that "the Godolphin mine was found very rich in copper close to the surface, and large profits were derived from it...The explorations were continued in more recent times by the aid of very powerful steam machinery below the points at which the lodes were found to produce copper and large quantities of tin were raised; but the low price of that metal and the heavy expense of draining the mines prevented their being profitable continued."

Henwood gave a detailed description of the lodes as they appeared at the time the mine was in operation. From this it is apparent that tin

Section 4: Levant to South Condurrow

was present in five out of the six lodes which were then being worked principally for copper. Henwood stated that in some places the tin was present in "large masses." Tradition adds that wolfram was also appearing in quantity in the bottom of the mine, the presence of that mineral usually being indicative of tin at greater depth.

From every point of view the deeper development of these lodes appears to offer great possibilities. The amount of water pumped in the past was admittedly heavy, but in view of the mine's situation beneath the Hayle River valley it is probable that much of this was surface water which could be prevented by suitable drainage arrangements from percolating into the mine at all.

In passing it is worthy of note that one of the small mines in this area, known as Wheal Osborne, is thought to be on a western continuation of one of the Godolphin lodes. An old friend, who was a shareholder in this small venture many years ago, told the writer that it contains "a large low-grade tin lode, worth about 1%" which, with the then price of tin, did not pay to work and was therefore abandoned.

(7) The Wheal Vor District

Situated on the eastern flank of Tregonning Hill is a tract of highly-mineralized killas, about 2 miles square, which gave rise to what were probably the richest tin mines ever discovered in Cornwall. The area contains numerous mines, but by far the most important are Wheal Vor and Wheal Metal.

The former property was worked on a large scale from about 1812 to 1847, by which time the workings, mostly on a single lode, extended from west to east for more than 6,600 ft. and reached a maximum depth of nearly 2,000 ft. It has been said that the mine "yielded at one period one-fifth of the entire tin produce of the world, making a profit of over half a million sterling." Unfortunately no record of production was preserved, but it must have come to a prodigious total. Hunt states in his "British Mining" that for a considerable period the output of Wheal Vor amounted to 200 tons of black tin per month, this being smelted on the mine. Incidentally a large amount of slag from the old smelter

Section 4: Levant to South Condurrow

has been cleaned up in recent years, much of which contained 36% of tin metal! Apparently absence of iron in the ore caused excessive slagging and made it a very difficult proposition for the old workers to handle.

The cessation of operations about 1847 was hastened by a costly lawsuit which had been dragging on for many years and which ultimately destroyed the enterprise. However, there is little doubt but that the original ore-shoot had been exhausted for when the mine was unwatered again at great cost in about 1854 it was found to be poor and was soon abandoned. The strange thing is that any company should have been willing to incur the cost of unwatering such an extensive mine without following it up with any worth-while development, especially further east which has always been thought to be a good prospect. As shown by the plans and sections very little cross-cutting was carried out and though other lodes were intersected little was done on them. The first working of the mine seems to have been a classical example of "picking out the eyes" without making any serious attempt to develop other ore-bodies. The 1854 working ended without doing any new development worthy of mention.

Realizing the possibilities still latent in the mine further eastward a new company was formed in 1906 to work the property and one of the first electrically-driven sinking pumps employed in Cornwall was used for the task. However, when the water had been lowered to about half the depth of the mine a mechanical failure wrecked the steam-driven generating set and the mine was consequently flooded again before any serious exploration could take place. Having had so much trouble with the plant and having uselessly expended much capital the shareholders were unwilling to go any further and the company was wound up. Thus ended the last and most ill-conceived attempt to rework Wheal Vor.

The company formed in 1854 to rework Wheal Vor and several other mines in the neighbourhood had conducted its affairs so recklessly that it was on the point of collapsing with the total loss of its capital. However, in the nick of time, an exceedingly rich discovery was made one-third of a mile south of old Wheal Vor, the new mine being named Wheal Metal. In an effort to retrieve something from the financial wreck the other mines were abandoned and all attention concentrated on the new property. This developed into another extraordinarily rich mine and, had the early profits from it not been swallowed up in the great losses incurred in the other mines of the group, it might well have been one

Section 4: Levant to South Condurrow

of the greatest prizes ever discovered in Cornwall. Unfortunately the company seems to have been interested solely in recouping its earlier losses and Wheal Metal, even more than Wheal Vor, was worked with reckless disregard for the future. Although other lodes are known to exist they were content to work two only, the plans showing that there was practically no cross-cutting done. In fact it seems that the company was only willing to develop the existing ore-shoots as long as the development would pay for itself.

Although Wheal Metal was a smaller mine than Wheal Vor it was of comparable richness. The tonnage mined was small, but it is said that for some years the ore crushed averaged 7% to 10% of black tin! From 1855 to the end of 1858 the monthly output of black tin averaged 40 tons and during the years 1865-69 about 60 tons per month. When Charles Thomas was writing in 1867 the returns were 70 tons to 80 tons per month. The mine continued to yield largely until the early '70's, when, as a consequence of a falling off of values and chronic lack of long-term development, the concern became unprofitable. In 1874 two eminent mining managers were called in to advise and they recommended certain lateral developments. However, the shareholders who had bought in at high prices while large dividends were being paid were unwilling to put up the capital required for this work and shortly afterwards all work was stopped and the property abandoned.

Whether further development eastward in Wheal Vor and Wheal Metal would result in the discovery of additional ore-shoots remains an open question. The views of several very capable mining men, who knew these mines intimately, are still preserved in old documents and from these it is clear that they thought that the chances of success were good. It is certain, however, that the belt of ground in which these mines are situated still possesses possibilities and that systematic cross-cutting in depth would be fully justified.

(8) The Porkellis Mine, Wendron

Situated almost in the centre of the great Carnmenellis granite mass there exists a group of tin lodes which have been worked at various periods in the Porkellis mine. This property has had a long though chequered history, having been many times re-opened and abandoned. The last working ended in 1938, when, in the midst of a slump in tin, the then sole owner abandoned the mine and sold the plant.

The section of the mine last in operation consisted of workings on two lodes, each of which had been developed for about 1,200 ft. in length and to a maximum depth of 560 ft. from surface. Of the two ore-bodies the "Old Men's" lode is worth about 1% of black tin over a width of 4 ft. to 5 ft. and the "Wheal Cock" lode approximately 30 lb. per ton over a similar width. The latter lode, however, varies greatly in width and value and some very rich bunches have been discovered in it. Through dilution of the ore from several causes the actual milling grade over a period of years was probably not much above 1% but the ore is cheap to mine and concentrate and over a number of years costs were only about 21s. per ton.

The two lodes in this part of the mine appear likely to join about 90 ft. below the present bottom workings. Although the other mines in this area have not persisted in depth and though there is evidence of the lower limit of the payable tin zone having been approached in Porkellis, it would be well worth sinking the mine a little further to investigate a possible junction of the two lodes. The best opportunities, however, are probably to be found in extending the developments on both lodes further eastward where their outcrops are known to exist but where no deep mining has been done. In addition there are other lodes, both to the north and south of this part of the mine, which certainly merit more attention than they have received in the past.

(9) Wheal Grenville and South Condurrow

The great central mining district of Cornwall, situated about Camborne and Illogan, is divided into two sections by the Carn Brea — Carn Entral granite ridge, which runs approximately east to west. The area south of this granite ridge is known as the "Flat Lode" district, the name being derived from a major lode which dips south at an unusually "flat" angle (on an average little more than 30°) and which has been worked in all

Section 4: Levant to South Condurrow

the numerous mines of the area. The Illogan district is also sub-divided by a major cross-course running north and south, this cross-course also being an extensive fault. It is the mines of the Flat Lode district, west of this cross-course, which it is now proposed to discuss. These mines — namely Wheal Grenville and South Condurrow — have between them produced over 31,000 tons of black tin but they still seem to possess possibilities.

In common with all the other Flat Lode mines these properties originally commenced on the numerous relatively-vertical lodes which abound in the area, but many years ago attention was diverted by the discovery of the celebrated Flat Lode. As has so often happened in Cornwall when a especially good lode has been discovered, this great ore-body was "flogged to death", all the other lodes being neglected and their development stopped. When ultimately the ore-shoots on the Great Flat Lode began to peter out at an inclined depth of about 2,400 ft., it was realized that much more attention should have been given to the other lodes. By this time, however, Grenville (which includes a part of South Condurrow) was in serious financial and man-power difficulties as a result of the 1914–18 war, which was then in progress and it was impossible to carry out the development which should have been done many years before. An attempt was made post-war to reconstruct the company to enable promising developments then taking place in the shallower levels to be continued, but this proved to be impossible and the concern collapsed in the great slump of 1920.

Mr Josiah Paull, for many years manager of the South Crofty mine, was also consulting engineer to the Grenville company in its closing stages. The writer discussed Grenville with him shortly before his death and he stated that he formed the highest opinion of the mine's possibilities at the time he was connected with it and greatly regretted its premature abandonment which was largely due to lack of foresight in earlier years. He remarked that at the time of closing the grade of ore at Grenville was 25 lb. of black tin per ton or at least 2 lb. higher than that prevailing at South Crofty, but whereas the latter was making a profit Grenville was making a heavy loss. This he attributed to the antiquated and inefficient machinery and milling plant, to lack of development for many years, and to heavy pumping costs, which could have been greatly reduced by extending one of the deep adits of the district into the property. Mr Paull, who prepared several reports on the property, stated that

Section 4: Levant to South Condurrow

it was responding wonderfully well to the limited amount of development which it was possible to do during the closing months of the concern's life. He did not recommend the unwatering of the deep portions of the mine at a future date as he regarded the ore-shoots on the Great Flat Lode as having been virtually "unbottomed". However, he considered that the chances of successful development of the several "vertical" lodes in the shallower levels — say, to a depth of 1,200 ft. — and also the possibilities of finding the Flat Lode westward of where it was lost by the faulting of a cross-course west of Marshall's shaft were very good.

The *Transactions of the Cornish Institute of Engineers* (Vol. X, 1925), contains a paper by Captain William Thomas, an authority on the Cornish mines, in which he argues that a large portion of the Flat Lode remains unworked in very shallow ground in the adjoining mine of South Condurrow. Indeed, looked upon as a shallow mining proposition — say, to a vertical depth of approximately 1,200 ft. — this district still appears to offer considerable possibilities.

Tolgus to Lambriggan

Section 4: Part II

TOLGUS TUNNEL TO LAMBRIGGAN

(10) Tolgus Tunnel

In 1919 the company operating East Pool and Agar formed a subsidiary to develop the eastern extension of its own lodes which at that time were most productive and profitable. The plan was to explore this eastern ground by means of a lengthy tunnel at the 255-fathom level (approximately 1,600 ft. from surface), which was to be driven eastward from the old workings in Wheal Agar. The tunnel was to have been 2,200 ft. long and from its eastern end it was intended to do the development necessary to determine the location of a new shaft in the old Tolgus mines.

The driving of the tunnel commenced in January, 1920, and when it had been extended 860 ft. a very fine north-dipping lode was encountered. This ore-body was probably close to the southern side of the tunnel for some distance before it was intersected, for it entered the drive at quite a narrow angle. Where exposed the lode averaged 76 lb. of black tin and 6 lb. of WO_3 per ton over a width of 8.5 ft. A cross-cut driven south to prove the full width of the lode disclosed a value of 148 lb. of black tin and 10 lb. of WO_3 per ton over the full lode width of 13 ft.

At the time of this important discovery the general opinion was that it was the lost eastern continuation of the Great Lode of the Agar and East Pool mines which, incidentally, is also the Main Lode of the South Crofty mine further west. This great ore-body has been extensively mined for a length of more than a mile and has yielded very considerable profits in all the mines worked upon it. However, a large feeder of water, amounting to about 240 gallons per minute, was cut with the lode in the tunnel and it was then found that the old flooded Barncoose mine was being rapidly drained. As the Tolgus tunnel is more than 600 ft. vertically below the deepest workings in Barncoose it is very probable that the tunnel lode has some connection with that worked in Barncoose.

Tolgus to Lambriggan

In passing, it is worthy of note that though the latter was originally a copper mine it changed to tin in depth. From the writings of Moissenet it is apparent that the mine was producing tin on a considerable scale during the '50s of the last century, which is evidence that the tunnel area is not too far east of the local emanative centre to yield tin in quantity, as has been suggested in some quarters.

Shortly after the discovery of the great lode in the tunnel the first of the misfortunes occurred that were to lead to its abandonment — namely, a serious pump breakdown at the Agar shaft on June 30, 1920. In order to prevent the mines from being flooded it was then necessary to erect a dam temporarily to seal off the additional water that had just been cut in the Tolgus tunnel. The dam was built near the eastern end and while the pump repairs were being carried out a borehole 410 ft. long was drilled northwards from a point a little west of the dam. Four additional lodes were thereby exposed and although the values in them were mostly low the core obtained from the lode passed through from 225 ft. to 228 ft. averaged 38.3 lb. of black tin per ton. Before it was possible to remove the dam the disastrous slump of 1921 had brought all operations to a standstill. Then, on May 18, 1921, the great "run" of ground occurred that destroyed both of the old East Pool shafts. Subsequently the company sank the new Taylor shaft further north, but the old workings, including the Tolgus tunnel, were never again unwatered.

In May, 1923, another new shaft was commenced by Tolgus Mines, Ltd., half a mile east of the end of the tunnel. The latter, incidentally, lies beneath a point about 100 ft. north-east of the junction of Chariot and Chili Roads. This second shaft was sunk in the expectation that the lodes of the western mines would continue productive of tin in depth on the eastern side of the Barncoose valley. Unfortunately the killas-granite junction proved to be very steep on the eastern side of the valley and instead of entering the granite the lodes dwindled in depth in beds of greenstone and became valueless. The enterprise was a complete failure and after doing a good deal of development at a depth of 2,000 ft. the mine was abandoned. It is easy to be wise after the event but subsequent developments suggest that the Tolgus shaft should have been sunk west of the valley and close to the end of the tunnel.

By 1942 the workings of the Taylor shaft section of the East Pool and Agar mines were becoming exhausted. In view of the then acute national need of tin and tungsten it was proposed to re-open the Tolgus tunnel area by driving a new tunnel from the Taylor shaft at a depth of 1,650 ft. This would have been 1,700 ft. long and would have cut the Great Lode about 100 ft. below the end of the old tunnel which, incidentally, is 1,533 ft. below the collar of the Taylor shaft. Unfortunately the national need of metal was so urgent that the Ministry of Supply, which at that time was in control, could not sanction any long-term developments and the exploration of the Tolgus Tunnel lode was again shelved. At the conclusion of hostilities the mine, which by then had become very poor and uneconomic, had to close. An attempt was made in 1946 to reconstruct the company with a view to resuming the development of the highly promising Tolgus Tunnel Section, but the Capital Issues Committee refused to sanction an issue of fresh capital and the company had to go into liquidation.

As can be seen from a plan of the district the end of the Tolgus tunnel is in the centre of a large area that is entirely unexplored in depth. Quite apart from the fine lode exposed in the tunnel and the indications in the borehole north there is a strong probability of other well-known lodes being intersected here in depth. As such this piece of ground contains interesting possibilities.

The Barncoose mine was not a wet one and it is probable that the water encountered in the Tolgus tunnel mostly came from that standing in the old workings. Once that mine has been drained there is no reason to anticipate that the working of the Tunnel lode would entail exceptional pumping costs. As, however, this ore-body may well prove to be the Great Lode on which there are immense workings further west, it is almost certain that those workings in East Pool and Agar will have to be unwatered to permit of the exploration of the lode to the west as well as to the east. Indeed the successful development of the Tunnel area would seem to be dependent upon it being worked in conjunction with South Crofty. That property already has to pump most of the water originating in the old East Pool and Agar mines and if it were decided to unwater the latter to the depth necessary to regain access to the Tolgus tunnel the overall pumping costs would not be very greatly increased. The Tunnel area could then be worked to a considerable depth without the necessity of installing any pumps in the East Pool and Agar mines and, if the venture were successful, the total pumping costs of these mines could be spread over a considerably larger tonnage of ore.

The old Agar shaft gives direct access to the Tolgus tunnel and, if it

Tolgus to Lambriggan

were unwatered from South Crofty and re-equipped, the exploration of the Tunnel area could be resumed immediately. Admittedly the shaft is in a bad condition and its reconditioning would be costly. Furthermore the eastern end of the Tunnel is over 1,600 ft. away and therefore the permanent working of the lodes of that area would probably not be economical from Agar shaft. Nevertheless such a plan would avoid the necessity of sinking a new shaft until the development of these lodes had progressed far enough to prove whether such an expenditure was justified.

(11) The Scorrier Mining Area

The Scorrier Area may be defined as a tract of country extending from the northern side of the town of Redruth to the village of Chacewater, a distance of about four miles; its width is approximately a mile and a half. The area contains over 20 mines, but, great as has been its past production, there can be no doubt that it is one of the best, if, indeed, not the most promising large mining area to be found in the whole of Cornwall.

The area is composed of killas situated on the northern flank of another granite intrusion — namely, that of Carn Marth. The metamorphic aureole of this intrusion extends over the whole area and, indeed, a good deal further north to a point around Porthtowan on the north coast where it links up with the aureole of the smaller intrusion at St Agnes Beacon. The Scorrier district contains many important lodes and numerous cross-courses; several elvan dykes pass through the area and one or two patches of greenstone are also in evidence. Indeed in many respects the area bears a striking similarity to its more famous sister district of Illogan and Camborne immediately to the westward.

The old mines here are nearly all shallow, very few of them having attained a depth of 1,000 ft. from surface, while most of them are considerably less. Although there has been an appreciable tin production from a few of these properties the district as a whole has hitherto been predominantly a copper one. The mines have mostly been worked on a small scale and succumbed during times of depressed metal prices or

Tolgus to Lambriggan

were abandoned by typically "copper" managements when the bottom of the copper zone had been reached. Very few of the mines reached the granite in depth, but it is significant that in several of them tin, arsenic, and wolfram were all beginning to make their appearance in some quantity. The writer considers that there is a striking parallel between this area and that of Illogan. In the case of the latter greater foresight and greater energy were responsible for the mines being sunk below the copper and into the tin zone, whereas development in the Scorrier district ceased at what was probably the transitional horizon.

Practically the whole of the mines in this area are connected to the Great County Adit, which drains over 30 square miles in this and the adjoining Gwennap copper district. Unfortunately this very extensive adit system is in a bad state of repair and is partially choked in places in consequence of which it is more of a liability than an asset to any future mining operations. With a view to obviating this difficulty the British (Non-Ferrous) Mining Corporation, Ltd., proposed in 1937 to drive a large new adit from the north coast, capable of serving all the mines of the area. This proposal formed part of an ambitious plan to develop the whole district on a large scale, but it was delayed for a while by the opposition of a single large mineral owner and the scheme ultimately lapsed.

An immense amount of data concerning this area and its mines has been collected recently and embodied in a set of reports; the information which these contain should be invaluable in any future large-scale planning of mining here.

(12) Stencoose and Mawla United

Immediately north of the Scorrier area and situated between it and the lesser copper-mining district of Porthtowan on the coast there are a number of lodes which have been but little explored. These veins occur over an area about $1\frac{1}{4}$ miles long by two-thirds of a mile wide and merit much greater attention than they have yet received. At the western end of this region there is a small mine known as Stencoose and Mawla United. In 1937 the writer collected a good deal of information about

Tolgus to Lambriggan

this property for a private syndicate who intended unwatering it, but for a variety of reasons the proposal did not come to fruition. The following notes are based on information obtained at that time.

The mine was worked by a small company during the period 1860-62. At that time old adit workings, 90 ft. from surface, were re-opened and a shaft sunk 60 ft. below adit level or 150 ft. from surface. Two narrow, north-dipping, stanniferous lodes occur at adit level, but it appears that the old workers were searching principally for copper ores. From their reports it is apparent that a lode, which was sometimes 5 ft. wide and carrying occasional bunches of copper ore, was being driven on below adit level. At this juncture what was considered to be a very promising tin lode was cut at the bottom of the mine, but it contained so much water that the small Cornish pump was completely overwhelmed, although the inflow does not seem to have been more than 300 gallons per minute and was probably less. To cope with this the pump was enlarged and an effort made to unwater the mine. Although information about subsequent events is inconclusive it is probable that the unwatering attempt was a failure and that the bottom of the mine was never seen again. It is known, however, that one of the proprietors made efforts to raise further capital but he was unsuccessful and the company was soon wound up. There is a persistent tradition in the locality concerning this discovery and the reason for its abandonment and when the matter was thoroughly investigated in 1937 it was found that the documentary evidence available fully confirmed the traditional story.

The mine is situated $1\frac{1}{2}$ miles north of the northern bastion of the Carn Marth granite mass and is in a direct line between that granite outcrop and that at St Agnes Beacon, $2\frac{1}{2}$ miles further north. The indications of metamorphism and mineralization throughout this region are suggestive of the existence of a subterranean granite ridge connecting the two outcrops at no great depth. Three prominent cross-courses also pass through the area and it is worthy of note that it was in their vicinity that the productive lodes of Wheal Charlotte, Wheal Towan, South Towan, Tywarnhaile, and Wheal Ellen were worked further northwards. Indeed Stencoose exists in an excellent "ore parallel" and it seems worthy of further investigation. In 1912 work was resumed on a small scale down as far as adit level, but, on the death of the principal backer, the mine was again abandoned.

This series of lodes undoubtedly persists over a considerable strike

Tolgus to Lambriggan

and on occasion rich stones of tin have been picked up in the fields a considerable distance east of the mine. Nearly a mile east of Stencoose, at another small property known as Wheal Concord, a surface discovery of tin was made some years ago, when the foundations for a hedge were being dug. This virgin lode was followed down for a few feet most profitably by the working man who had discovered it, but, lacking means, he was unable to continue its exploration any further. Indeed Stencoose and the greater part of this area is still practically virgin and seems well worth thorough investigation.

(13) Wheal Coates, St Agnes

The numerous lodes at Wheal Coates cut through the small granite intrusion on the western side of St Agnes Beacon, but the later and more important workings on the Towanrath lode are situated in the killas between the granite and the cliffs and even extend some way out beneath the beach. The landward part of the mine, which is marked by extensive openworks, was sunk to a depth of 450 ft. which is not far below sea-level, but little is known of the very old workings. The important part of the mine is that near to the cliffs which was being worked in the '70's and '80's of last century. From this section about 700 tons of black tin were produced, but work was ultimately suspended through the prevailing low price of the metal. The machinery was left intact for a while as it was hoped to re-open the mine, but later the whole of the plant was dismantled. In 1911 the property was unwatered and partly equipped, but as in the case of Stencoose, on the death of the principal backer in 1914 it was again abandoned without any further development.

The workings on the Towanrath lode extend 480 ft. below adit (30 ft. above sea-level) and for a maximum distance of 950 ft. along the strike. With the exception of the bottom level, which is only driven for 160 ft., each succeeding level in depth is longer than the one above. The greater part of the ground developed has been stoped away and the length of the ore-shoot (judging from the area stoped) is likewise increasing with depth. The lode, which dips south at about 80°, varies in width from 2 to 12 ft., but in addition to the lode proper the enclosing rock is full of little veins of clay and cassiterite extending for a considerable distance from the lode. The whole mass constitutes an ore-body of exceptional width, the stopes varying from 8 ft. to 25 ft. and sometimes as much as 40 ft. or even more. It is also worthy of note that the average width, as well as the length stoped, was increasing with depth. Notwithstanding the exceptional size of the stopes and the general absence of pillars the walls are stated to be strong and firm.

When the mine was re-opened it was found that the low-grade ore left by the old workers only averaged 18 lb. to 20 lb. of black tin per ton, but throughout the life of the mine the average is stated to have varied from 1% to $1\frac{1}{4}$%. Given further development there seems no reason why that average should not be maintained in future. In addition the ore is very cheap to break and crush and easy to concentrate as it contains few sulphides. During the last working the price paid for stoping (by hand drilling) was only 2s. 6d. to 3s. per ton and the pilot milling plant, using Cornish stamps, crushed over $4\frac{1}{4}$ tons per head per day.

Approximately 100 ft. south of the Towanrath lode another orebody known as the Copper lode has been intersected by cross-cuts at the two deepest levels and driven on for a maximum length of 400 ft. This lode averages 4 ft. to 6 ft. in width and in the 70-fathom western drive, under the beach, contains good "grey copper" (chalcocite). In addition it carries tin of considerably higher value than that found in the Towanrath lode; a small stope below the 70-fathom level is said to average 50 lb. of black tin per ton. The manager during the last working is of the opinion that the Copper lode is in reality only a branch of the main or Towanrath lode and he thinks that it will join the latter in depth. He states that the development points on the Towanrath lode were suspended in the old working wherever the lode had become small or had split up into stringers in proximity to the cross-courses. He considers that if development were resumed there is every likelihood of the lode continuing to be productive both laterally and in depth.

At the 80-fathom or bottom level the old workers drove a crosscut about 300 ft. south of the Copper lode in the hope of cutting the celebrated West Kitty lode. Nothing of value was intersected in this cross-cut, but a great deal of water was encountered. The last company built a dam in the cross-cut thereby greatly reducing the quantity to be pumped, which subsequently amounted to less than 200 gallons per minute. The mine is served by a compound shaft, perpendicular to the 60-fathom level below adit and sunk below that horizon on the Towanrath lode to the 80-fathom or 480-ft. level.

The probability of the considerable extension both in strike and depth of the Towanrath lode, the great width of that ore-body, the cheapness with which it can be mined and dressed, and the possibility of intersecting other lodes by further cross-cutting, all combine to make Wheal Coates worthy of serious consideration.

(14) Wheal Kitty and West Wheal Kitty
St Agnes

Situated on the north coast, about three miles due north of the Scorrier area, is to be found the small but once productive tin district of St Agnes, whose numerous mines are all worked in the killas. Granite is in evidence in small outcrops to the east and west, but it has never yet been encountered in any of these mines. These properties are characterized by the extraordinarily "flat" northern dip of their lodes and also by the very complex faulting that exists. The recorded output of black tin from the two largest mines (Wheal Kitty and West Wheal Kitty) is some 30,000 tons between 1853 and 1930. It is known that there were other considerable sales before 1853, but no records exist. In common with most of the other mines of the area these properties appear to be worked out with the exception of one highly-important ore-body, but that is so rich that it merits much further attention.

Owing to the financial and man-power difficulties resulting from the 1914–18 war these mines were both abandoned between 1916 and 1918. In 1926 they were combined and re-opened in the expectation of being able to rediscover the great ore-shoot of the Wheal Kitty lode that had been lost through faulting. After initial disappointments the company succeeded in locating the ore-body and thereafter they did 2,596 ft. of development on it, the average for the whole of this footage being 80.8 lb. of black tin per ton over a lode width of 28.4 in. The richest part of the ore-shoot persisted for 900 ft. in length and the inclined depth over which it was exposed was 250 ft. The values in a sump winze, the deepest point at which it was seen, were not quite as high but even there they still averaged 53.5 lb. over a lode width of 28.3 in. The writer can testify having personally examined the workings, that there

Tolgus to Lambriggan

was little indication, even at the deepest part, of any deterioration of the general productivity of the lode. Indeed, he is of the opinion that further development would have shown that this rich ore-body extended much further in depth. The milling recovery of the whole of the ore obtained from this lode, both from the great shoot and from the lower grade ground further east and west, averaged 39.8 lb. of black tin per ton; the output finally attained was about 46 tons of black tin per month. Unfortunately the price of tin, which had averaged £291 per ton in the year in which the mine was restarted, continued to fall steadily until, in September, 1930, all operations had to be suspended. By that time the price had receded to £133 per ton.

In 1937 a new company, under the title of Polberro Tin, Ltd., was formed to take over the assets of the defunct Wheal Kitty company and to continue, under the same management, the deeper exploration of the great lode that had had to be abandoned in 1930. For a variety of reasons it was considered that this could best be accomplished by re-opening the Turnavore shaft of the old Polberro mine, to the north-west, and sinking it to the requisite depth necessary to intersect the rich Wheal Kitty lode.

The work was subsequently carried out but it failed in its primary object. The shallower and much lower-grade West Kitty lode was, indeed, intersected by the shaft at a depth of about 730 ft. and a considerable amount of development and stoping on it was done, but, on the whole, it was a disappointment. The writer examined the workings on several occasions and there would seem to be a strong probability that the lode for which the company was searching was in some way connected with the numerous small veins and branches that were cut in the shaft at a depth of about 1,000 ft. Had the finance been available to drive eastward on one or more of these little veins at the 1,020-ft. level there is every possibility that they would ultimately have coalesced to form the great ore-shoot which was being sought. Indeed, it is worthy of note that it was by following similar "strings" that the older Wheal Kitty company had discovered the great ore-body 10 years previously.

Unfortunately financial difficulties, which were entirely due to the recent war, crippled the Polberro company and prevented it from raising further funds. All the spade work had been done and an independent engineer reported favourably on the prospects, recommending that a further £20,000 be provided to enable development to be continued. This the directors were unable to raise. The ore in sight on the top lode

Tolgus to Lambriggan

was rapidly being stoped out and an appeal was made to the Government to provide financial assistance to enable further deep development to be carried out. This assistance, however, was refused and, in consequence, operations had to be suspended in March, 1941, the plant subsequently being dismantled and sold.

That so much work should have been done and that everything should have had to be abandoned when success might soon have been achieved was a great pity. Undoubtedly the possibilities latent in this rich lode merit a further effort being made to locate it from the Turnavore shaft at some future date.

(15) The Perranporth Mines

A mile and a half east of the best tin mines of St Agnes a small but very interesting patch of granite outcrops in the cliffs at Cligga Head. For a little over half a mile west and a mile and a half east of this outcrop the aureole of metamorphism is very marked for a distance of over half a mile from the cliffs. The small intrusion at Cligga Head is apparently a portion of a very much larger mass to the north-east which has now been eroded by the sea. Even under the land itself it has been proved that the granite is far more extensive than is evident at surface for, in the course of mining operations, it was intersected a considerable distance east of Cligga Head.

Extending throughout the whole length of the strip of metamorphosed killas and running parallel to its long axis are a number of lodes which have yielded a great amount of copper and a little tin; there are also two important elvans present. The principal mines worked on the lodes were, from west to east, Wheal Prudence, Perran St George, Perran United, and Wheal Leisure. The deepest workings are in the last-named mine, under the town of Perranporth itself, but even there they only extend to about 700 ft. below sea-level, whereas the majority of the workings are only 500 ft. or less below sea-level. The statistics of production are very incomplete, for they only extend back to 1824, whereas it is known that some of the mines were working as long ago as 1600. The total recorded output of copper ore is 177,000 tons.

Tolgus to Lambriggan

The collapse of mining in this district about the middle of the last century was due not so much to the exhaustion of the mines as to the very heavy cost of pumping and to a series of disputes between the companies working the various mines. The only possible means of working these properties economically is to combine them into one major concern, when, in spite of the water, they might be an economic proposition. Old reports suggest that some appreciable reserves of copper still exist here, especially in the Wheal Leisure part. This is possible, but the writer considers that the possibility of tin occurring beneath these masses of copper is of far greater importance. Whether deeper sinking would expose a tin zone in or close to the granite is an open question, but from the general geology of the area and the position of the lodes relative to the granite they seem to be ideally situated for such a change taking place in depth.

In this connexion the multitude of small tin/wolfram veins which occur in the patch of granite at Cligga Head (and in the killas immediately adjacent to it) is not without interest. The veins at Cligga, which have been worked from time immemorial, were mined on a much larger scale during the years 1938 to 1944. These operations, which extended down to sea-level, demonstrated that the workable ore-bodies occur in both granite and killas but were limited to within a radius of 100 ft. to 200 ft. of the vertical junction between the two rocks; at a greater distance from the contact they are too small or too poor. Whereas at surface the veins are all very narrow, at sea-level, 300 ft. below, some of them have opened out to form well-defined if small lodes 12 in. to 18 in. wide. In the deepest workings a few of them are carrying either good tin values or massive wolfram and arsenopyrite in a quartz gangue.

When examining these workings the writer came to the conclusion that although uneconomic as so far exposed, the veins merit deeper exploration — say, to at least 500 ft. or 600 ft. below sea-level — which would probably also provide useful information concerning the possibilities of tin occurring beneath the copper in the lodes of the adjacent killas country.

(16) The Great Perran Iron Lode

Two miles north of Perranporth and at the northern end of Perran Bay a remarkable vein, known as the Great Perran Iron Lode, outcrops in the cliffs. This great lode in its course eastward at first strikes about 35° south of east, but further inland it assumes a more nearly due west-to-east direction. It has been explored in several mines to shallow depths, but nowhere exceeding 420 ft. The workings on it extend for upwards of $3\frac{1}{2}$ miles from the coast, but it is said to be traceable much further east. Its mean dip is about 45° south and although the width is very variable it is unusually large, varying from 3 ft. to 50 ft. or even more.

Portions of the lode have yielded large quantities of blende and, where crossed by some of the numerous cross-courses which intersect it, lead and copper have also been mined. Its principal yield, however, has hitherto consisted of iron ores. Of these the main one in very shallow levels is brown haematite. Collins states that beneath the haematite "sometimes only a few fathoms from surface and always before sea-level is reached, the lode appears to be largely composed of spathose carbonate of iron (with occasional broad belts of dark compact blende and much more rarely veinlets of yellow copper ore or galena). On the whole the carbonate of iron is most abundant in the foot-wall, the blende in the central or upper parts."

The potentialities in depth of this lode have long been a matter of contention. Most present-day geologists argue that it is in effect a cross-course or, at least, an iron-bearing vein of later age than the tin and copper lodes and, as such, has no possibility whatsoever of yielding those metals in depth. Collins, a Cornish mining geologist of unique experience, who had an intimate knowledge of the lode, was, on the contrary, strongly of the opinion that the mass of iron and other minerals which it exhibits is merely the iron "cap" or gossan indicator of an "exceedingly rich" copper vein at greater depth. The possibilities latent in Collins' suggestion are so great that the matter ought to be tested by a comprehensive series of drill-holes put down to intersect the lode at depths of at least 1,500 ft. Even if the lode failed to yield copper in depth its great possibilities for blende should not be overlooked in view of the increasing scarcity of zinc.

(17) The Shepherds Lead Mines, Newlyn East

In the vicinity and south of Shepherds Station, on the branch line from Chacewater to Newquay are situated the surface remains of the celebrated Shepherds Mines. These were discovered accidentally by Sir Christopher Hawkins in 1816 during the draining of a marsh and subsequently proved very productive, the lead being smelted on the mines and a good deal of silver recovered in the process.

The property consists of virtually two separate mines, the northern being the first discovered, but this was only worked to a depth of approximately 360 ft. Sir Christopher Hawkins later found another an even more productive group of lodes about a third of a mile further south and these were eventually worked to a depth of 750 ft. In the mistaken belief that the latter workings had been prematurely abandoned they were re-opened in 1881-6, but were found to be almost completely exhausted. An old man believed to be the only miner now living who worked in Shepherds during this period states that the mine was un-watered to the bottom. It was found that the various development points had been abandoned by the old workers when they had become poor and the new company, finding them unproductive, did practically no new development and soon abandoned the mine again. The small production of lead during the last working was obtained from the old stopes etc.

At the time of the 1881 flotation great stress was laid on the possibilities of these lodes being developed in the virgin ground to the west of the mines, but very little was done to explore these possibilities. It is interesting that the farmer working the land immediately to the west of the mine states that, if ploughing deeply, he can always turn up stones of galena. His son confirms this and, from what the two men have told the writer, it would appear that the lead has been found approximately on the line of strike of the main lode on the southern part of the Shepherds mine. These indications of lead are in a large field immediately south of the railway line and about 700 ft. west of the dumps of the old south mine. The field is on relatively-high ground, but it is usually wet and marshy which is strongly suggestive of the presence of the outcrop of one of the typically wet lead lodes of the district.

Old plans show that there were seven lodes in these mines which

Tolgus to Lambriggan

were thought to extend westward into the virgin ground of which the field mentioned forms a part. This unexplored ground is in the centre of the best "lead country" of Cornwall and constitutes one of the very few areas where there seems to be a reasonable likelihood of making further lead discoveries of importance. Although geophysical prospecting, as so far tried in Cornwall, has not yielded any very positive results, it is an interesting speculation whether it would prove of value in a survey of this piece of land. The area certainly merits a careful examination.

(18) Silverwell Lead Prospect, Mithian

A further possibility of discovering lead has been brought to the writer's notice recently and this is at Silverwell, which is to be found on the 6-in. Ordnance map of Cornwall, Sheet LVII, N.W. The location of the supposed lode is at Lat. 50° 17' 30" and it is said to be traceable from approximately Long. 5° 9' 30" to at least 1,200 ft. west of that point. Near a well, marked on the Ordnance map as "Silver Well" water pipes were being laid in a trench a short while before the recent war and it is said that splendid little stones of galena and "sugary" quartz were found in the trench. Many years ago lead was discovered in a field west of the shallow valley which lies west of the well. Furthermore a farmer, who some years since was draining the marsh land at the bottom of the valley, also cut a lead lode approximately in line with the two former outcrops. The consensus of local opinion is that an east-to-west lead lode of some importance exists here, but nothing has been done to investigate it. At about the time of the discovery of lead near the well another trench for pipes was being dug up through the valley (in a north to south direction) and a great rock of galena "as large as a horse's head" was found some way north of the supposed east to west lode. No other sign of lead could be found in the trench, but this very large stone of galena is suggestive of a shoad-stone having been carried down the valley from a lode at Silverwell.

The difficulty in connexion with the development of this lode is that the local authority obtains a supply of water from the well. Any mining would be almost certain to interfere with this, but it might be possible to come to some arrangement with the authority for an alternative source of water supply.

(19) Lambriggan Zinc and Lead Mine, Perranzabuloe

The principal shaft of the mine at Lambriggan is situated at the eastern side of a steep valley, $1\frac{3}{4}$ miles south of Perranporth Station and two-thirds of a mile east of Mithian Halt (on the branch line from Chacewater to Newquay). About a century ago the property was worked under the title of South Saint George, when much zinc blende was found in the shallow levels. In 1927 the mine was re-opened for lead and zinc, but the latter was found to be predominant. A considerable amount of development was done during this period, but the mine never reached the milling stage and was precipitately abandoned in 1930 in consequence of the heavy fall in metal prices that occurred at that time.

Although there are at least four east-to-west lodes and two striking north to south in the property it was found during the last working that the "old man's" workings were confined almost entirely to a single east-to-west lode which dipped south. The main shaft proved to be perpendicular and sunk to a depth of about 300 ft. Connected with it there was an adit and three levels, their respective depths from surface being 42 ft., 165 ft., 230 ft., and 295 ft. The last-mentioned level only consisted of a cross-cut driven far enough south to intersect the lode. During the 1927–30 period the levels at 165 ft. and 230 ft. were described as the Nos. 1 and 2 respectively and were considerably extended, principally eastward. In addition the shaft was sunk about 100 ft. and a new level (No. 3) opened out at a depth of 400 ft. from surface. The lode was explored for a length of about 1,350 ft. and, although the values were patchy, it was proved that there were at least two well-defined ore-shoots, the one in the eastern levels being over 400 ft. long and that west of the shaft at least 150 ft. in length. The company found that the previous workers had mined most of the lead, leaving the zinc *in situ*, but in the old parts of the mine and also in the newer developments the blende was predominant, galena only occurring in quantity in a few places. From

Tolgus to Lambriggan

available records it appears that the ore-shoots sampled by the late company averaged about 17% zinc (metal) and 6.5% lead (metal) over an average lode width of 4 ft. A little silver was also present with both the zinc and lead.

Near the surface the lode was small and poor, but the width and values increased with increasing depth to the horizon of the No. 2 level. At the No. 3 level 400 ft. from surface the ore-body was still wider, sometimes being as much as 8 ft. in width, but, although explored for a length of 650 ft., the values at that depth were disappointing. However, encouraged by the occasional high-grade bands of ore in this, the bottom level, a sump winze was commenced east of the shaft. Within 15 ft. of sinking the lode had improved considerably in appearance and at the time of the abandonment it gave grounds for thinking that further good ore-shoots would have been discovered at greater depth.

A geophysical survey of the property was also made and confirmed that other lodes were present; two of them subsequently exposed on surface contained both lead and zinc. Beyond some inconclusive cross-cutting at the No. 3 level no attempt was made to intersect any of these other lodes in depth and in any future reworking of the property it is obvious that much further attention should be paid to them.

After examining the mine down to the 295-ft. level the writer was much impressed with its possibilities for zinc. In view of the greatly changed world outlook for that metal since 1930 it certainly seems that the mine is worthy of much more extensive trial, both laterally and in depth. A certain amount of lead may also be expected, but the property should be regarded primarily as a zinc producer as is evident from the preponderance of blende in the dumps around the main shaft. Any future reworking of the mine should be greatly facilitated by the development carried out by the late company, in addition to which electric power is now available within 770 yd. of the main shaft. As far as water is concerned the influx in 1928 was at first heavy, but after the adits had been reconditioned it decreased to a maximum winter flow of 286 gallons per minute, which is not at all excessive for one of these typically wet lead-zinc mines of the Perranzabuloe district.

Section 4: St Austell to Trebartha–Lemarne

Section 4: St Austell to Trebartha–Lemarne

Section 4: Part III

ST.AUSTELL–PAR MINES TO

TREBARTHA–LEMARNE

(20) The St Austell-Par Mines

South of the Hensbarrow granite mass and situated immediately to the east of St Austell there is a most important mining district which still possesses great possibilities. The area in question extends from St Austell to Par Station, a distance of $3\frac{1}{2}$ miles and from the coast inland for about $1\frac{1}{2}$ miles. Many famous tin and copper mines occur within this stretch of ground, the recorded production of which amounts to approximately 23,000 tons of black tin and 372,000 tons of copper ore.

Most of the numerous lodes of this area have a bearing somewhat S. of E. and N. of W. and thus lie roughly parallel to the margin of the granite. The country consists of killas, much of which lies outside of the metamorphic aureole. The granite probably exists deeper to the south-east than in the north-western part of the area and it is significant that whereas the eastern mines were predominantly copper producers those at the western extremity yielded mainly tin. As the ore zones within the lodes dip in the same direction as the killas-granite contact it is very probable that an unworked tin zone exists beneath the exhausted copper deposits of the mines at the southern and eastern portions of the area. Indeed, Collins remarked of the Crinnis mine: "There is probably no old copper mine in the county which better deserves to be sunk in search of tin ore than this".

It is significant that Par Consols, which was the deepest and also the most easterly of these mines, began to yield tin after reaching a depth of 250 fathoms. Par Consols was the creation of the famous J. T. Treffry, but his successors lacked the vision and energy that he displayed in all his great enterprises. Consequently when, after his death, the copper deposits were becoming exhausted the mines were abandoned in 1869 on the advice of "copper" managers who did not see their way to provide

Section 4: St Austell to Trebartha–Lemarne

the necessary machinery for extending the mines in depth and providing them with the requisite new crushing and dressing plant.

The northern part of Par Consols is known as "Puckey's North Mine." This section lies closer to the granite than the principal workings and although it had only reached a depth of 160 fathoms it is reputed to have produced tin in considerable quantities as well as copper. The sales of black tin from Par Consols are recorded as having been 3,785 tons. About a mile to the west the same run of lodes yielded in the Wheal Eliza group of mines over 11,000 tons of black tin. The latter properties were very profitable and were only abandoned in the great depression of the 'nineties.

Officers of the Geological Survey remark of the St Austell area: "In nearly every respect the geological conditions are comparable with those of the mining district of Camborne and Redruth." Undoubtedly the conditions prevailing there are favourable for the deposition of tin and one cannot but help thinking that there is scope for a great deal more development between St Austell and Par. This applies especially to deeper sinking.

(21) Holmbush, Kelly Bray, and Redmoor Mines, Stoke Climsland

This extensive group of mines is situated in the killas on the western flank of the Kit Hill granite boss and about one mile north of Callington. Although the earliest workings here commenced before the 19th Century the mines are still relatively shallow, their depth in fathoms being as follows: Holmbush 190, Kelly Bray 110, and Redmoor 125.

In Holmbush, the most northerly of the three mines, there are four or five east-and-west lodes, but only two, the north-dipping Holmbush and Flop Jack lodes, have been explored in depth. During the 19th Century the mine produced large quantities of copper ore and arsenical pyrites. Kelly Bray was worked on a single south-dipping lode, hitherto principally a copper producer. At Redmoor, the southern mine of the group, there are several east-and-west lodes. In the northern part of that mine there is a north dipper that has been worked on a small scale for

Section 4: St Austell to Trebartha-Lemarne

tin and tungsten during the present century. South of that is Johnson's or the main lode, which dips south and varies in width from 3 ft. to 5 ft. Still further south is the north-dipping Vivian's lode, which comes in contact with Johnson's lode at the 125-fathom, or bottom level and, finally, the Great South tin lode, which, dipping north, is thought likely to junction with Johnson's lode at a depth of about 150 fathoms. Both Vivian's and the Great South tin lodes are 5 ft. to 6 ft. in width. Intersecting the east-to-west lodes of all three mines there is a north-and-south lead lode, which is reputed to have produced a considerable tonnage of lead in the earlier workings. Incidentally, this crossing of east-to-west tin-copper-arsenic lodes by a productive north-to-south lead lode is an extremely rare occurrence in Cornwall. The workings on the lead lode connect the Holmbush and Redmoor mines down to a depth of 112 fathoms and, therefore, when dealing with the water of one mine it is necessary to handle that of all three. The total incoming quantity is 500 gallons per minute, most of which comes from Holmbush.

After sundry vicissitudes and periods of inaction the whole group was re-opened by the Callington United Mines Ltd., in 1888, and worked as one concern until the end of 1892. The company was, however, short of working capital from the very commencement of operations and it was entirely due to this cause that the mines were again abandoned. Extensive and detailed reports on the properties were prepared during this last period of activity by J. H. Collins, Capt. F. Wright, and Capt. George F. Richards (the then Agent for the Duchy of Cornwall). These reports, which are still preserved, make it clear that the mines possessed great possibilities, but that they could not be worked profitably on the restricted hand-to-mouth policy then in vogue and which made adequate development impossible. In general the reports agree that Holmbush was changing from copper to tin in depth and that, by reason of the position of the granite, the tin zone was likely to be encountered first at the eastern end of the mine. It was thought that the unexplored parallel lodes merited serious attention and that there was still a possibility of opening up a good deal of payable ore on the lead lode, both in Holmbush and at Redmoor. The opinion was expressed that Kelly Bray, like Holmbush, was turning to tin in depth and should be actively developed. At Redmoor Collins reported stopes running from 1% to 2% of black tin per ton and, among major developments that were recommended, both he and Capt. Richards were emphatic in stating that the main shaft should be

Section 4: St Austell to Trebartha–Lemarne

sunk sufficiently deep to investigate the junction of the Johnson's and Great South tin lodes. Both men thought that Redmoor would develop into an important tin producer in depth and that, in all three mines of the group a great deal of arsenic and probably some further copper ores would be discovered.

The money for the necessary developments and modernization of the plant was, however, not forthcoming at that time of exceptionally low metal prices and, after a feeble and unsuccessful attempt at reconstruction, the whole venture was finally abandoned about the end of 1894. As already stated, one of the northern lodes in Redmoor was worked profitably about 1913 for tin and wolfram, but only to the depth of adit level. Since then there have been some further small operations and one or two rather unfortunate flotations in connexion with the property. No serious work, however, has been attempted at any of these mines since the early 'nineties and this rich square mile of ground still merits development in depth.

(22) The Prince of Wales Mine, near Callington

The Prince of Wales mine is about $2\frac{1}{2}$ miles east of Callington and situated on the southern side of the main road from Callington to Gunnislake. It has been worked near the southern margin of the metamorphic aureole surrounding the granite masses of Kit Hill and Hingston Down, which, at no great depth, probably unite to form a continuous ridge. Located thus at the foot of a granite hill and parallel with its long axis the property is ideally situated, as was pointed out by Charles Thomas as long ago as 1867. The mine was an old one when it was re-opened in 1862, but with the exception of short periods of suspension it was thereafter worked continuously on a modest scale until its abandonment about 1914. During the period 1864–1908 the recorded sales totalled 11,340 tons of copper ore, 1,031 tons of black tin, 8,153 tons of arsenical pyrites, 7,720 tons of iron pyrites, and a certain amount of lead and silver. From beginning to end, however, operations were conducted on a hand-to-mouth basis, the lack of finance always making adequate development impossible. This is especially true of the French company which

Section 4: St Austell to Trebartha–Lemarne

assumed control in 1899 and whose ultimate failure resulted in the final closing of the mine. In the writer's opinion the property still possesses great possibilities; indeed, it would seem to be one of the best prospects in East Cornwall. It should be recorded that much of the information which follows was given to the writer by the late Mr. James A. Michell, who was manager of the mine until a short while before its abandonment.

The property possesses several lodes having an average bearing of a few degrees north of east. The most important of these, from north to south are as follows: The North Prince of Wales, Good Luck and George lodes. Further south there are the Well and Harrowbarrow lodes; the latter carries high-grade arsenical and silver ores and was worked for silver in the adjoining Queen mine. The principal ore-body is the steeply south-dipping Prince of Wales lode, which has been worked to a depth of 180 fathoms and for a maximum length on the strike of about 1,900 ft. Most of the mine's copper output came from this lode, but in depth it yielded principally tin, the average for the whole of the ore stoped probably being rather over 1% of black tin. The width of the lode varies from 2 ft. to 6 ft. During Mr Michell's management stopes at the 135-fathom eastern level were sometimes worth as much as 50 lb. of black tin per ton over a width of 3 ft. and others above the 155-fathom level 40 lb. over 5 ft. In this connexion the late Mr. Josiah Paull, who, at the time in question was manager of the neighbouring Hingston Down mine, told the writer that when he went underground at Prince of Wales he was greatly impressed by the strength and character of the main lode there. From the 155- to 180-fathom levels the lode was 5 ft. to 6 ft. wide and often very rich, but at the 180 it was abruptly cut off by a south-dipping "slide." Under the previous management the main shaft was sunk to the 193-fathom level and, assuming that the slide had resulted in a normal fault, cross-cuts were put out north at the two bottom levels in the expectation of again intersecting the main lode. It seems, however, that the lode which was cut was actually the North lode. There are grounds for thinking that the slide may have given rise to a reverse fault and that, in reality, the Prince of Wales lode exists south of the shaft below the 180 and not to the north as had been assumed. Whatever the explanation this strong and well-defined lode was cut off sharply at the 180-fathom level and has not been seen below that horizon. A further effort to locate its downward continuation is obviously warranted.

Little work has been done on the North lode, but some of the lodes in

Section 4: St Austell to Trebartha-Lemarne

the southern part of the property have been extensively worked down to the deep adit, which, at the point where it intersects the Good Luck lode, is 42 fathoms from surface. Incidentally, had this adit been extended about 900 ft. further north it would have come into the workings on the Prince of Wales lode at a depth of about 55 fathoms from surface and would thereby have considerably lessened the amount of water to be pumped from shallow workings in winter time. Furthermore, if the 55-fathom cross-cut north had been further extended (as a continuation of the deep adit), it would have tested the ground between the Prince of Wales and Hingston Down mines and would have unwatered the latter's lodes (some of which are but little explored) to a depth of upwards of 90 fathoms from surface. It should be mentioned that the water pumped at Prince of Wales is only about 200 gallons per minute, but the main shaft (Watson's) is small and bad, being only about 11 ft. by 6 ft. and very crooked below the 90-fathom level.

The Good Luck lode is a well-defined ore-body, 4 ft. to 5 ft. in width and said by the old tributers, who had a good opinion of it to be "tinny". It was worked to some extent down to deep adit level. Its outcrop is 1,000 ft. south of that of the Prince of Wales lode, but it dips north, whereas the latter dips steeply south. A cross-cut was therefore driven south at the 166-fathom level from the workings on the Prince of Wales lode in the expectation of cutting the Good Luck lode at no very great distance. Although the cross-cut was extended more than 400 ft. nothing of importance was cut and, as considerable water had been encountered, further exploration was abandoned and a dam erected 190 ft. south in the cross-cut. There would seem to be scope for a more extensive search for this lode below adit level.

The George lode, which lies nearly 1,500 ft. south of the Prince of Wales lode, dips steeply to the south and has been extensively worked down to the deep adit, which is there 40 fathoms from surface. The vein is 3 ft. to 4 ft. wide and although it carries some copper its principal content is very rich arsenical pyrites. The late Capt. George Richards, Mineral Agent to the Duchy of Cornwall, had great faith in this lode and he often speculated as to what its contents would prove to be in depth beneath such great quantities of arsenic.

The Prince of Wales lease is about one mile long by two-thirds of a mile wide, but even this piece of ground has as yet been very imperfectly explored, while all around it there exists a great extent of entirely

Section 4: St Austell to Trebartha-Lemarne

undeveloped country. Indeed the existing workings are little more than the nucleus of the development necessary for the opening up of such an extensive and promising tract of land. It is a matter of speculation whether granite will be encountered at greater depth, but it is obvious that at the modest depths attained on the main lode that ore-body is already well within the stanniferous zone. Furthermore there seems no reason for thinking that the deepest workings are anywhere near the lowest limits of that zone and the prospects for the re-discovery of the Prince of Wales lode below the 180-fathom level are definitely good. The possibilities of much deeper mining, of a great extension along the strike of the main lode, of developing the other lodes in depth, and of cutting further ore-bodies by extended cross-cutting all combine to make this mine a most attractive prospect.

The property is worthy of a far more extensive trial than it has yet received.

(23) Trebartha-Lemarne Mine, North Hill

This small but interesting property, which was for long forgotten, has recently been re-discovered by Sir Arthur Russell to whom the writer is indebted for many of the following particulars. The mine, which is situated in the grounds of Trebartha Hall, 8 miles due north of Liskeard, is on the killas-granite contact at the eastern margin of the Bodmin Moor granite mass. Small workings here, which are said to be three or four centuries old, yielded a little gold, from which a ring was made for the Rodd family, whose seat for many centuries was Trebartha Hall. The 19th Century workings, however, came about through the discovery of old records at the Hall by a local doctor who was the then Mr. Rodd's friend. A small "cost book," or unlimited liability, company was subsequently formed to develop the mine and Mr Rodd became one of the shareholders.

The plans show that there are three or four lodes. The "Gulley" or southernmost, which was worked open-cast, was mined to adit level and consisted of bunches and strings of mineral, all of which paid to mill. The workings below adit, however, were all confined to the main

Section 4: St Austell to Trebartha-Lemarne

lode and consisted of two shafts, an adit (about 10 fathoms deep at the western shaft), and a level 10 fathoms below adit. The bottom level was driven for a distance of 430 ft. and some of the ground above it, as well as above the adit, was stoped. Rodd's, or the western shaft, was sunk 10 fathoms below the bottom level.

The lode on which these workings occur is well-defined, with good walls and said to be from 4 ft. to 6 ft. wide. Its strike is 25° S. of W. and the dip N. 75°. The workings are mostly in killas and rather wet, but at the western end the granite was encountered and the levels became entirely dry. In the killas the ore contained principally arsenic and wolfram. On entering the granite the lode was at first pinched, split up into strings, and hard, but on penetrating further into the granite the ore-body improved and is reputed to have looked most promising for tin. A small dump of arsenical-wolfram-tin stuff near the western shaft is said to have come from these western developments in the granite. Incidentally a recent sample of the small tailings dump gave a value of about 30 lb. of mixed black tin and wolfram per ton.

The mine was worked on a small scale for a few years during the 'eighties. The power for pumping, winding, and the stamping and dressing of the ores was all obtained from a 30-ft. diamater water wheel with a breast of 6 ft. During this period the following sales of ore were reported: 20 tons of black tin, 18 tons of iron pyrites, 201 tons of crude and refined arsenic, and 14 tons of tungstate of soda. At first the concentrates were taken to a works near Callington for calcination and the extraction of the wolfram. Later the company erected its own arsenic works, but the expenditure thereon strained its very limited resources. A few dividends were paid, but it then became necessary to raise further capital, which most of the shareholders, being merely working men on the estate, were unable to do. Furthermore Mr. Rodd was opposed to the continuance of the mine, which was becoming much larger than he had foreseen. As it was not far from the Hall the site of the mine, the noise of the stamps at night, the fumes from the arsenic burning and, above all, the poisoning of the fish in the Trebartha ponds through the discharge of so much arsenical water were all very objectionable to him. The mine was therefore closed in about 1888 and trees were planted to obscure the site.

Since the death of the late Major Rodd the estate has been sold and the celebrated mansion demolished. Therefore, apart from the possibility

Section 4: St Austell to Trebartha-Lemarne

of trouble with the Fishery Authorities over the pollution of the River Lynher, the major obstacles to the re-working of the mine have been removed. Geologically the property is in a good position, being situated as it is on the contact. In addition the hillside rises for 400 ft. above the adit and, if the lodes persist westward into the granite, a great deal of mining would be possible without having to do any pumping. Very little cross-cutting has been done and there are possibilities of other lodes being discovered in the vicinity. Indeed from every point of view this small but very promising mine is worthy of a much more extensive trial than it has yet received, both laterally and in depth.

Section 4: St Austell to Trebartha–Lemarne

Index of Mine Names

Alfred Consols	105-107, 122
Ball Dees	132
Balleswidden	61-62
Balmynheer	140-141
Barncoose	163-165
Basset and Grylls	132-138
Billia	68
Binner Downs	7, 117-120, 122-123
Birch Tor	38
Bog	75
Boscean	60
Boswin	139-140
Botallack	57-61, 100, 149
Breage	100
Callington United	183-184
Calvadnack	141-143
Cardrew	6
Carnyorth	57
Carzise	114-115, 117
Cligga Head	8, 173-174
Clowance	114
Combellack	131
Crenver and Wheal Abraham	7, 117-124
Crinnis	181
Croft Gothal	85-86
Crowan Consols	123
Devon Great Consols	42
Ding Dong	62-63
Dolcoath	41, 43, 121
Drym	123-124
Duffield	111
Durlo	68
East Blue Hills	26
East Charlotte	26

East Pool	22, 43-44
East Pool and Agar	163-166
East Rosewarne	112
East Wheal Darlington	75
East Wheal Grylls	87
East Wheal Lovell	128, 131-132
East Wheal Rodney	76, 79-83, 153-154
East Wheal Rose	122
Fatwork	131
Garlidna	138-139
Geevor	56-57, 61, 119, 148-149
Giew	10-11, 67-71, 151-152
Godolphin	7, 92, 100-102, 115, 155-156
Great County Adit	25, 167
Great North Downs	6, 27
Great Perran Iron Lode	11-14, 175
Great Western Mines	87
Great Wheal Alfred	105-108, 122
Great Wheal Fortune, Breage	99-100, 105
Great Wheal Fortune, Marazion	83-85
Great Wheal Grylls	87
Great Wheal Vor United	10, 93, 95
Great Work	91, 97-99, 115
Grylls Wheal Florence	87
Guskus	85
Gwallon	83
Gwennap United Mines	24
Gwinear Consols	112
Gwyn and Singer	100
Halamanning	85-86
Hallenbeagle	6
Herland	105, 108-109
Hingston Down	185-186
Holmbush	182-184
Jantar	134
Kelly Bray	182-184
Lambo	114
Lambriggan	178-179

Leeds and Saint Aubyn	87
Levant	8-9, 52-62, 147-150
Lewis	116
Marazion Mines	75-76, 85
Mellanear	42, 107-108
Mengearne	131
Millpool	87
Mount Wellington	22-23, 25
New Rosewarne	112
North Levant	56
North Rosewarne	112
North Treskerby	6
Oatfield	119
Old Treskerby	6
Old Trumpet	129
Old Wheal Lovell	130-131
Old Wheal Prosper	85
Owen Vean	76-77, 82, 85, 153-154
Par Consols	181-182
Parbola	112-113
Peevor	6-7
Penberthy Crofts	76, 84-85
Penhale Wheal Vor	100
Perran Saint George	173-174
Perran United	173-174
Polberro	27-28, 45, 172-173
Polcrebo Downs	124
Poldown	10, 95
Polengrean	139
Polhigey	141-143
Polladras Downs	100
Polrose	100
Porkellis	23-24, 27, 132-138, 143, 158-159
Prince of Wales	184-187
Prosper United	85
Providence	66-67
Puckey's	182
Queen	185

Redmoor	182-184
Reeth Consols	10-11, 67-69, 151-152
Relistian	109-110, 112
Retallack	85-86
Rosewall Hill and Ransom United	66, 150
Rosewarne and Herland United	112
Rosewarne Consols	112
Rosewarne United	112
Roskear	44
Rospeath	83-85
Ruby Iron Mine	14
Saint Aubyn and Grylls	87
Saint Ives Consols	7, 64-66, 68, 150-152
Shepherds	176-177
Silverwell Lead Prospect	177-178
South Condurrow	159-161
South Crenver	117, 120, 124
South Crofty	44, 77, 119, 122, 160, 163, 165-166
South Providence	10-11, 68
South Rosewarne	112-113
South Saint George	178
South Towan	168
South Wheal Speed	68
Stencoose and Mawla United	167-169
The Lovell	130-131
The Weeth	111
Tindene	86
Tolgus	42, 44
Tolgus Tunnel	163-166
Tolvadden	86
Trebartha Lemarne	26, 187-189
Trebarvah	87
Tregembo	117
Tregonebris	131
Tregurtha Downs	74, 76-83, 85, 122, 153-154
Treleigh Wood	6
Trencrom	70
Trenear	132

Trenoweth	117, 120
Tresavean	42
Trevarthen Downs	85
Trevascus	110-112
Trevenen and Tremenheere	130
Trumpet Consols	129-131
Trumpet United	129
Tywarnhaile	168
Unity Wood	44
Wendron Consols	132
Wendron United	139-140
West Alfred Consols	107
West Godolphin	7, 102, 116, 155
West Poldice	44
West Rodney	153-154
West Rosewarne	104, 112
West Treasury	114
West Wheal Darlington	75
West Wheal Grylls	87
West Wheal Kitty	9, 170-173
West Wheal Providence	114
Wheal Alfred Consols	86-87
Wheal Ann, Marazion	85
Wheal Ann, Wendron	129, 131
Wheal Bellan	149
Wheal Bolton	83-85
Wheal Boxer	87
Wheal Breage	98
Wheal Briggan	6
Wheal Buller	42
Wheal Busy	6
Wheal Castle	87
Wheal Chance	6
Wheal Charlotte	168
Wheal Chippendale	75-76
Wheal Coates	9-10, 169-171
Wheal Cock	57, 59-60
Wheal Concord	169

Wheal Courtis	123-124
Wheal Crab	75-76, 153-154
Wheal Darlington	153-154
Wheal Drea	61
Wheal Dream	129
Wheal Dumpling	123
Wheal Eliza	182
Wheal Ellen	168
Wheal Enys	139
Wheal Friendship	85
Wheal Grenville	159-161
Wheal Grylls	87
Wheal Gurlyn	116-117
Wheal Hampton	27, 76-77, 79-83, 116, 153-154
Wheal Harmony	6
Wheal Hartley	111
Wheal Hermon	149
Wheal Jennings	112-113
Wheal Julia	123
Wheal Kitty, Polpeor	70
Wheal Kitty, Saint Agnes	9, 26-27, 171-173
Wheal Leisure	173-174
Wheal Lemon	87
Wheal Margaret	70
Wheal Margery	66
Wheal Mary	70
Wheal Metal	10, 94-97, 99, 156-158
Wheal Neptune	86-87
Wheal Noon	129
Wheal Osborne	115-116, 156
Wheal Owles	60-61
Wheal Prosper	85
Wheal Prudence	173-174
Wheal Prussia	6
Wheal Puffet	139-140
Wheal Racer	11, 66, 150
Wheal Reeth, Breage	22-23, 98-99
Wheal Reeth, Saint Ives	67-68

Wheal Rodney	27, 76-77, 79-83
Wheal Rose	6
Wheal Ruby	139
Wheal Sarah	119
Wheal Sisters	70
Wheal Speedwell	87
Wheal Strawberry	118, 123
Wheal Towan	168
Wheal Trannack	129
Wheal Treasury	114
Wheal Tremayne	114, 117
Wheal Trenwith	7, 65-66, 71, 150-151
Wheal Trewavas	100
Wheal Valls	129
Wheal Vernon	139
Wheal Virgin	75-76, 82-83, 153-154
Wheal Vor	10, 45, 91-97, 99-100, 115, 156-158
Wheal Widden	129
Wherry	88-89
Whiteworks	23-25